Interior Designs I

室内空间设计 1

《室内空间设计》编写组编　李婵译

辽宁科学技术出版社

图书在版编目（CIP）数据

室内空间设计.1/《室内空间设计》编写组编；李婵译.
—沈阳：辽宁科学技术出版社，2015.3
ISBN 978-7-5381-9071-7

Ⅰ.①室… Ⅱ.①室…②李… Ⅲ.①室内装饰设计
—作品集—世界—现代 Ⅳ.① TU238

中国版本图书馆CIP数据核字(2015)第024518号

出版发行：辽宁科学技术出版社
　　　　　（地址：沈阳市和平区十一纬路29号　邮编：110003）
印　刷　者：沈阳天择彩色广告印刷股份有限公司
经　销　者：各地新华书店
幅面尺寸：225 mm×300mm
印　　张：22
字　　数：100千字
出版时间：2015年3月第1版
印刷时间：2015年3月第1次印刷
责任编辑：鄢　格
封面设计：何　萍
版式设计：迟　海

书　　号：ISBN 978-7-5381-9071-7
定　　价：180.00元

联系电话：024—23284360
邮购热线：024—23284502
E-mail:lnkjc@126.com
http://www.lnkj.com.cn
本书网址：www.lnkj.cn/uri.sh/9071

CONTENTS 目录

News 设计新闻
- Office in Santiago — 圣地亚哥办公设计等 — 2

They Say 设计专题
- Hotel Murano — 穆拉诺酒店 — 4
- Chiswick Moran Hotel — 克里斯莫兰酒店 — 10
- Hilton Niseko Village — 希尔顿二世古山庄 — 14
- Superbude Hotel — 超休闲酒店 — 20
- Vincent Hotel — 文森特酒店 — 24
- Hotel Riva — 河岸酒店 — 30
- Haymarket Hotel — 海玛特酒店 — 36
- Hotel Skt Petri — 第一佩奇酒店 — 42
- The Grand Daddy Interesting — 老爸趣味酒店 — 48
- Marriott Marquis — 万豪伯爵酒店 — 54

Projects 设计案例
- Peter Restaurant — 彼得餐厅 — 60
- Barolo Ristorante — 巴罗洛餐厅 — 64
- Sushi Restaurant — 寿司餐厅 — 70
- Restaurant Mangold — 曼古德餐厅 — 74
- Pan-American Restaurant — 全美洲餐厅 — 78
- Fremantle Media — 弗里曼特尔传媒办公大楼 — 82
- Papsa Showroom — 帕帕莎陈列室 — 88
- Ecomplexx Officez — Ecom公司的综合办公大楼 — 94
- Toyota Tsusho — 丰田通商 — 98
- Wieden+ Kennedy — 文登+肯尼迪公司办公大楼 — 104
- 980 5th Avenue — 第五大道980号 — 110
- Tiburon House — 蒂布伦住宅小屋 — 116
- Private Residents House — 私人住宅 — 122
- Vader House — 维达住宅 — 128
- Shuwai — 墅外 — 134
- The World's First Underwater Spa — 世界首个水下疗养地 — 140
- Cidade Jardim Mall — 西大德·加帝购物商场 — 144
- Imall Coloure — 怡玛伊考拉沙龙 — 150
- Tribeca — 迪比克俱乐部 — 154
- Honda Scooter Store — 本田小型摩托车商店 — 160

Interview 设计交流
- The personal sense is the roof of inspiation ——An interview with Kelly Hoppen — 以直觉激发灵感 ——访凯丽·赫本 — 166

Review 设计评论
- Focal Point in a Sea of Colours ——Colourway in Hotels — 对焦缤纷色彩 ——谈酒店色彩运用 — 170
- Peculiar Colour Design ——An analysis of Hotel Riva Design — 别样的色彩设计 ——河岸酒店设计浅析 — 172

Book Review 书评
- Cross the Border of Design — 跨越设计的边界 — 174
- Capture refined elegance, create cozy space — 营造格致高雅，获取惬意空间 — 176

Antonia Maio and Javier Quinteiro of O Antídoto have designed an office interior in Santiago de Compostela, Spain.
The installation consists of 800sq.m over two floors: a ground floor for offices, and a basement for social uses of the company.

O Antídoto建筑公司的设计师Antonia Maio和Javier Quinteiro共同完成了位于西班牙圣地亚哥–德孔波斯特拉市的办公设计。

Interior stylist and designer Faye Toogood has completed an installation to display shoes for fashion brand Comme de Garcons.
A series of stacking building blocks made from plaster were designed to help re-configure and change the way the space is used. The purity of the white plaster contrasts against the roughness of the rendered concrete walls and a series of interlocking copper pipes provided a strong industrial element to the design as well as providing plinths for the more practical purpose of displaying muli-coloured sneakers.

室内设计师Faye Toogood完成了时尚品牌Comme de Garcons的鞋店设计。

Zaha Hadid Architects recently completed a temporary chamber music hall at the Manchester Art Gallery, just in time for the second annual Manchester International Festival.
The 82-by-56-foot hall, which remains open for public viewing through September 1, was specifically designed to house solo performances of compositions by Johann Sebastian Bach during last month's festival. It features a translucent fabric ribbon swirling throughout the space, intended to create a spatial and visual response to the relationships of Bach's harmonies.

扎哈·哈迪德建筑公司近日及时的为第二届曼彻斯特国际艺术节完成设计了当代室内音乐厅，该项目坐落在曼彻斯特艺术展览馆内。

Japanese Designers-Nendo have completed a shop front and interior for Italian clothing company ASOBIO in Shanghai, China.
This is ASOBIO's first shop, a spacious bi-level interior with a generous opening. The shop's theme is 'focus', so the designer positioned monotone photographs on the floor and walls, and varied the size to imitate the effect of a camera's zoom lens, and the sharpness to recall the sense of being out of focus.

日本设计公司Nendo 完成了意大利服装品牌ASOBIO上海店的设计。（Nendo http://www.nendo.jp/en）

Russian architect Peter Kostelov has refurbished an apartment in Moscow using industrial materials.
Earlier it used to be a studio apartment; practically it was a one bedroom apartment. After reconstruction there came out a few more separate rooms. All these were placed around the space of 86 sq. m.
The open space with a column in the centre was filled with a frame made of steel tubes. The frame hasn't been trimmed with any decorative materials except of metal cover, which can be easily replaced by wood, stone, paper wall, cart or whatever.

俄罗斯设计师Peter Kostelov完成了一个位于莫斯科的公寓翻修，装饰运用了工业元素。（Peter Kostelov http://www.kostelov.ru，pk71@ya.ru）

Japanese Designers-Case-Real, headed by Koichi Futatsumata, have completed a boutique interior in Fukuoka, Japan.
"The double curve"—The shop spaces formed with two curves. One big curve expands obliquely into the inside considering the view from a street in front and the movement line and another gentle curve of the ceiling link in three dimensions.

日本设计公司Case-Real，在设计师Koichi Futatsumata的带领下完成了一个位于日本福冈的时尚用品店的室内设计。（CASE-REAL http://www.casereal.com）

Berlin Architects—J Mayer H have completed a permanent exhibition on sustainability for car brand Volkswagen's Autostadt visitor attraction at their factory in Wolfsburg, Germany.
The exhibition LEVEL GREEN was opened on the 4th of June 2009 and encompasses approximately 1000sq.m. The design was executed by the use of easily processed wood composite sheets (MDF) with varying thickness according to the structural and geometrical demands.

柏林建筑公司J Mayer H完成了他们在德国沃尔夫斯堡的一个关于大众汽车品牌旅游景点的可持续性的长期展览。（J Mayer H http://www.jmayerh.de）

Romanian Architects—A.A. Studio have completed the refurbishment of a private apartment in Bucharest.
We present you a generous space, of almost 400 sq.m , which resulted from uniting two apartments placed in mirror. Regarding the functional distribution of the spaces, the architects came up with a simple solution2: the day area placed in one apartment, and the night spaces in the other apartment, united by the fitness room. All these spaces are united by the furniture design and the lighting.

罗马建筑公司AAstudio完成了位于布加勒斯特的私人公寓翻新。

Reykjavik designer Sruli Recht has completed the interior of his own flagship store in an abandoned fishery in Reykjavik, Iceland.
Everything used was reclaimed from the now abandoned construction sites around reykjavík – from the dried and weather worn-shipping palettes to the long wood scaffolding, old metal frames, steps and wheeled bases.
The high walls are lined with a patterned corrugated cardboard, and the industrial castors were bestowed by the custodial facility from their inoperative waste containers.

雷克雅未克设计公司Sruli Recht完成了他自己位于冰岛雷克雅未克的旗舰店的室内设计。（Sruli Recht http://www.srulirecht.com）

Interior Architects—i29 and architecture firm Eckhardt & Leeuwenstein have designed a series of boardrooms for an investment group in Amsterdam.
All three boardrooms and a lounge are executed in an overall design concept. Large round lampshades spray painted gold on the inside, seem to cast light and shadow oval marks throughout the whole space. These ovals define the separate working areas.
The lounge area has, in combination with the white marble flooring these same light/shadow patterns that cover the bar and benches in silver fabrics.

室内设计公司i29和建筑公司Eckhardt & Leeuwenstein共同为一个在阿姆斯特丹投资集团设计了一系列办公空间。（http://www.i29.nl）

Design consultancy JHP was appointed by O2 to work on the design of a new store concept for the telecoms retailer. Decidedly moving away from being a 'phone shop' to being a 'platform for total connectivity', the new concept had to truly evolve the format of retailing communication devices. The intangibility of O2's content and services should be core to the store of the future.

O2公司指定JHP设计公司进行下一阶段的店面设计，将有更多店面升级为JHP概念。

Franklin Azzi Architecture has participated in the competition for the redesign of "le Palais des Beaux-arts" this July. This world-renowned historical monument, situated on the Champs Elysées, near the Seine and les Invalides, was originally realised for the world exposition in Paris in 1990. After being closed for ten years, it is reopened in 2005. Recently a competition was held for its renovation, and Franklin Azzi is working on a fabulous proposal for it.

法国富兰克林Azzi建筑事务所2009年6月参加巴黎大皇宫国家美术馆项目竞赛。

Hotel Murano

穆拉诺酒店

Given the commission to renovate this weathered chain hotel, the design team looked to the flourishing local art community for influence. In the same manner, the design team wanted to link the hotel to the community using glass as its vehicle.

Taking the hotel lobby back to its original, pure architecture offered a harmonious environment for art glass installations. It was critical that the backdrop be minimal and neutral to allow the art to be the focus.

Art glass is incorporated into the architecture—the front desk, entry doors, lobby chandelier and public restroom sinks were all created by internationally known artists. A cool blue glow floods the entry and lobby bar through stacked glass walls while the bar itself has a slump glass counter, illuminated from within.

Each of the 21 guest floors is dedicated to a single artist, featuring work displayed behind a customised etched glass wall engraved with artist quotes and commentary. Photographs and sketches along the corridor and in the guest room shed light on the artistic process.

Hotel Murano's rooms are individually unique while sharing an extensive attention to detail. Rooms feature hand-blown glass bedside lamps and glass-topped vanities alongside custom-designed furniture.

我们接到的任务是将一座饱经风霜的连锁酒店翻新。设计小组从当地盛行的艺术社团那得到很多灵感。同样，设计小组想将酒店和艺术联系起来，用玻璃艺术做直通车。

原有的酒店大堂，纯净的建筑风格为安装玻璃艺术品提供了一个和谐的氛围。但是其背景幕很小，并且为中性色彩，急需让艺术成为此处的焦点。

玻璃艺术品与建筑融为一体，摆在酒店的前台、入口的门、大堂吊灯及公共卫生间的水槽处的艺术品都是国际知名艺术大师的作品。凉爽的蓝色光洒满了入口和大堂酒吧，并透过层叠玻璃墙。吧台处也装饰有垂直的、彩色条纹玻璃，照明安装在吧台里面。

客房共21层，每层都由一位艺术家的作品来装饰。作品展示在预制的刻字玻璃墙内，墙上展示艺术家的留言和解说。沿着走廊，还有客房，都布置了照片和素描，强调了艺术的氛围。

穆拉诺酒店的客房与众不同，并注重细节。客房内，那手工吹制的玻璃床头灯和玻璃面的洗手盆，还有定做的家具使客房独具特色。

Location
Portland, USA
DESIGNERS
Corso Staicoff, Inc.
COMPLETION
2008
PHOTOGRAPHERS
Jeremy Bittermann, David Phelps, Dan Tyrpak

项目地点
美国,波特兰市
设计师
科索斯泰科夫股份有限公司
完工时间
2008
摄影师
约翰·克拉克,大卫·菲尔普斯

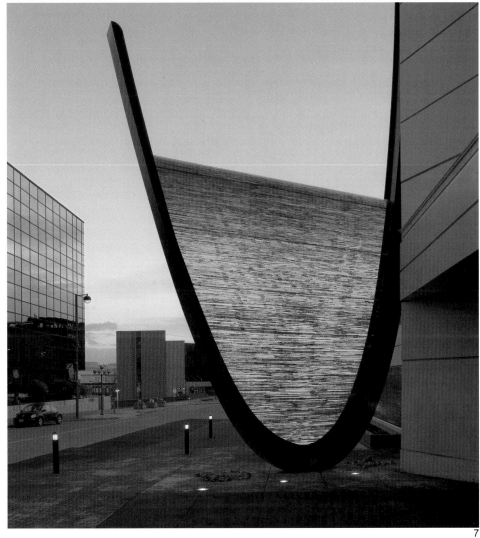

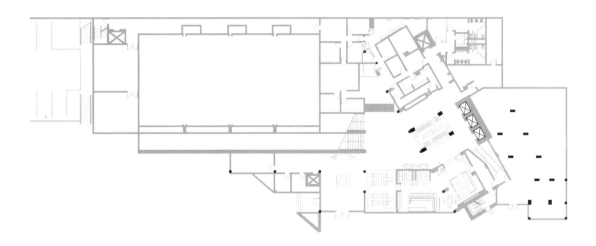

Chiswick Moran Hotel

克里斯莫兰酒店

This 120-bedroom hotel is situated on Chiswick High Road in a vibrant quarter of south west London. Housed in a 1960s' former office building, the concept for the hotel is 'West Coast/West London'. The vibe is a contemporary evocation of 60s' California, a link between this happening area of London and the glamour of LA.

The hotel is announced by tropical palms and a sweeping porte-cochere, beneath a cantilevered concrete canopy. A palette of heavily veined marble and stained oak boarding unites a flowing sequence of lobby, bar and restaurant. The Globe Bar takes its name from the huge shimmering globe that pivots between reception.

The restaurant goes by the name 'Napa,' synonymous with great quality and laid back style. The striking feature is series of screens of polished stainless steel and rotating green Plexiglas ellipses affording glimpses into the resident's bar, a tucked away corner of moody smoked mirrors and cow hide upholstery.

All have a full wall of glazing dressed with colour washed voiles and graphic black and white patterned curtains. Bathrooms are characterised by profiled ceramic tiles to an original 1960s' design in tangerine, lime and slate grey. The fittings are similarly unconventional; the basin and WC are triangular.

拥有120间客房的克里斯莫兰酒店坐落在伦敦东南部的克里斯公路上,那是一个很有活力的街区。这座建筑原是19世纪60年代的办公楼。酒店的设计理念是"西海岸(西伦敦)",风格是对19世纪60年代的加州风格建筑的现代诠释。它也是现实的伦敦和辉煌的洛杉矶的结合体。

酒店的特色在热带棕榈和悬挑的混凝土雨篷下快捷的车辆通道中彰显出来。深色木纹的大理石和涂漆的橡木板将大堂、酒吧和餐厅联系在一起。环球酒吧因悬挂于接待处和酒吧之间那个闪光的、旋转的大球而命名。

餐厅名为纳巴(美国加利福尼亚州西部一城市,位于奥克兰以北,是纳巴山谷的中心,此山区是有名的葡萄园地区),象征着高品质和轻松的氛围。更引人注目的地方是一面由多个椭圆形磨光不锈钢片及旋转的绿色有机玻璃片组合的屏风。透过屏风可以瞥见酒吧座位区—— 一个布置了模糊的烟熏镜和牛皮座椅的隐秘座位区。

所有客房都是落地窗,用巴里纱(一种轻而透明的薄纱)窗帘和黑、白图案的窗帘点缀。卫生间采用了19世纪60年代设计的橘红色、绿色、石灰色的铸型瓷砖贴面。这里的设施也与众不同,手盆和座便都是三角形的。

Location
London, UK
DESIGNERS
Project Orange
COMPLETION
2006
PHOTOGRAPHERS
Gareth Gardner

项目地点
英国,伦敦
设计师
奥林奇设计
完工时间
2006
摄影师
加雷斯·加德纳

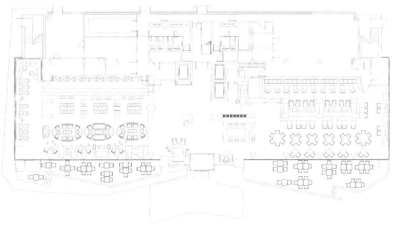

Hilton Niseko Village

希尔顿二世古山庄

Niseko is the place where the guest can enjoy its dynamic nature. Excellent powder snow during the wintertime brings lots of skiers from all over the world. Hilton Niseko Village aimed to create warm, welcoming resort hotel for those international guests.

In the lobby lounge, there is the huge fireplace hanging from the ceiling; furniture surround that fire. The designers try to create the space as if the guests having the warm and relaxing time at the mountain hut. Ezo pub offers a great "après" venue overlooking the Niseko Village, featuring private karaoke rooms, refreshing drinks and a snack menu.

二世古区的自然风光不断变化，让游客欣赏。冬季洁白、优质的雪吸引来世界各地的滑雪爱好者。希尔顿二世古山庄的设计目标就是成为一个吸引这些来自世界各地游客的度假型酒店。

在大堂休息区，有一个大型的壁炉从顶棚一直悬挂下来，家具围绕壁炉摆放。设计师试图设计这样一个空间：客人仿佛置身于山中温暖的小木屋里，放松身心地度假。Ezo酒馆是一个很好的集会地，那里俯瞰二世古山庄，设有私密的卡拉OK间，提供新鲜的饮料和小吃。

Location
Hokkaido, Japan
DESIGNERS
Hashimoto Yukio Design Studio
COMPLETION
2008
PHOTOGRAPHERS
Nacasa and Partners inc

项目地点
日本,北海道
设计师
桥本由纪夫设计工作室
完工时间
2008
摄影师
桥本由纪夫设计工作室有限公司

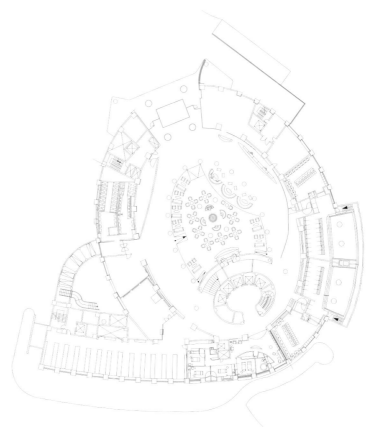

Superbude Hotel

超休闲酒店

In autumn 2006 Armin Fischer was assigned the task to develop a hotel for backpackers. The building, an old printing plant, is located between two main traffic arteries on the outskirts of Hamburg's St. Georg district.

Four sections of the building were interconnected. Four staircases and two elevators bring guests to their rooms. All public spaces had to be accommodated in the basement. The majority of the interior furnishing is composed of banal everyday objects and materials, which reflect the "Hamburg" theme. Buckets, for instance, were used to create an oversized fountain along an exterior wall. Pallets serve as frames for the couches, Beer crates from a well-known brewery in Hamburg were upholstered with genuine leather and transformed into stools. The lounge can also be used for meetings. A projector and a dolby surround-sound system give presentations audiovisual substance.

Stackable beds enable the 74-room hostel (148 stackable beds) to provide space for a total of 240 guests—from backpacker to business traveler: with its creative concept Superbude attracts a diverse mix of guests. Superbude is also a great place for young families with one or two children because the extra beds are already included in the room.

2006年秋天，阿明·费舍尔接到一个项目——为背包旅游者们设计一家酒店。该建筑原是一家印刷厂，它坐落在两个主要交通道之间，位于汉堡郊外的圣乔治区。

建筑的四个部分互相连接。四个楼梯和两部电梯将客人领到房间。所有的公共空间都容纳在地下一层。室内装饰主要由日常的用品和材料做成，反映汉堡的生活主题。例如，沿着外墙悬挂的吊桶被用来造一个大喷泉，货盘被当成长沙发的框架。知名的汉堡酿酒厂啤酒箱围上真皮，变为座椅。休闲区还可以作为会议厅。投影仪和一个立体声音响系统为演示提供了视听设备。

74间客房内放置了148张可折叠的床，可以为240位客人提供住宿，从背包旅行者到商人，酒店以它创新的设计理念吸引了不同的客户群。酒店还是年轻父母带上一两个孩子入住的好地方，因为客房内早已为孩子们准备了加床。

Location
Hamburg, Germany
DESIGNERS
Armin Fischer
COMPLETION
2008
PHOTOGRAPHERS
Eckhart Matthäus

项目地点
德国·汉堡
设计师
阿明·费舍尔
完工时间
2008
摄影师
克哈特·马修

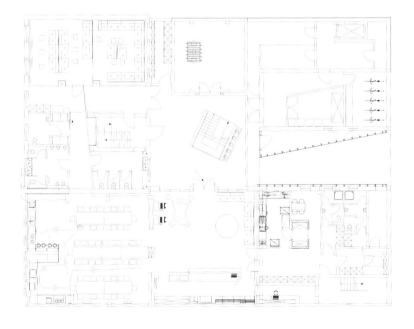

Vincent Hotel

文森特酒店

Located on Lord Street, the Vincent has the prime location in town and one of the finest in Northwest England. Southport is just a few hours by train from London.

Six floors of understated luxury, including a boutique hotel with 60 guest studios, residences and suites, a destination with café-deli, the V-Spa and gym. Dark woods abound, and neutrals with accent colours give an extra edge of luxury to the overall feel of the hotel.

The design of the Vincent follows clean, modern lines. The Vincent's classic-chic design starts the moment you enter the large, dramatic Welcome area with highly reflective black ceiling and limestone "feel tile", which gives the space a feel of elegance and glamour. The ground floor private bar boasts an ultra-luxurious feel, with a unique golden mosaic and high gloss wall panelling. The ground floor Café Deli blends modern furnishings, dark woods and amber tones.

The 225-square-foot V-Residence is a stylish and contemporary room with some really unusual extras, including a feature artwork screen which doubles as a personalised window drape. The individually-designed guest rooms grow in style and comfort through the V-Studio and V-Corner Studio on to the Vincent's luxurious Penthouse.

文森特酒店坐落在勋爵街，那里是市区中的一个好位置。它也是英格兰西北部最好的酒店之一。南港离伦敦只有几个小时火车里程。

时尚的文森特酒店有六层，整体装饰是一种不过分的奢华风格。酒店有客房、套房、工作室共60间，并且有咖啡厅、水疗和健身的场所。丰富的黑色木和中性设计之中的强调色给人一种高档酒店的感受。

酒店的设计使用了简洁、现代的线条。文森特酒店传统而又别致的设计感受从你走进门的那一刻开始，大尺度的、惹人注目的接待空间内，顶部是高挑的、反射光的黑色顶棚，周边是石灰石瓦，一种高雅和迷人感油然而生。位于一层的私密酒吧更是一个让客人感到特别舒适的好地方，装饰有独特的金色马赛克、高抛光的墙面板。一层的咖啡美食厅运用了现代的家具、黑色的木和琥珀色调。

20.9平方米的客房是一个时尚、现代的所在，一些真正不凡的附加装饰，如特别的艺术品布幔也是个性化的窗帘让人惊喜。个性化设计的客房风格不同，很舒适。顶层的豪华客房设计高档。

Location
Southport, UK
DESIGNERS
Design LSM
COMPLETION
2008
PHOTOGRAPHERS
Design LSM

项目地点
英国，南港
设计师
LSM设计公司
完工时间
2008
摄影师
LSM设计公司

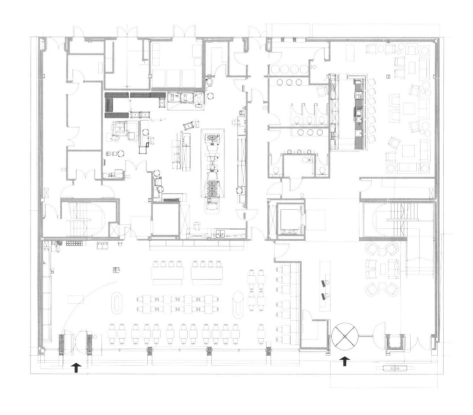

Hotel Riva

河岸酒店

The Croatian Island of Hvar is often regarded as the Monte Carlo of the Adriatic. It is very popular with the rich and famous attracting deluxe yachts that fill the Hvar harbour.

Built in the medieval style, the hotel's original construction, including fluted columns and stone walls, is frequently exposed. The hotel is entered from an extensive terrace positioned on the waters edge. The historic fabric is juxtaposed with trendy, colourful furniture and fixtures intended to appeal to the young, or young at heart, clientele. The hotel includes bespoke wallpaper decorated with large-scale drawings of nudes specially created for the project by Jestico + Whiles. The attention-grabbing wall treatment and other slightly risqué design elements, help to reposition the hotel as an alluring and desirable destination for the modern jet-set.

Once inside the bedrooms, the design is brave and uncomplicated. Blocks of red are combined with images of vintage film stars screen printed onto cotton fabric backdrops to the beds. A local stone was sourced for expanses of wall cladding combined with beige sandstone and slate flooring and a limited colour palate of red and black for the furnishings. The bathroom sinks are also red. The glass wall between the bathroom and bedroom gives a greater sense of space, with no doors and just etched stripes on the glass to give some element of privacy.

In addition to the 45 rooms and nine suites, Jestico + Whiles has designed a new hotel bar with a backlit, green, etched glass counter and a teak veneered top. A fusion restaurant is located on the other side of reception with further seating on the enlarged terrace.

克罗地亚的赫瓦尔岛通常被看作是亚德里亚海的蒙特卡洛（汽车品牌）。这里受有钱人士的青睐，知名的、迷人的豪华游艇在赫瓦尔岛上随处可见。

该建筑是中世纪时期的风格，酒店原有的框架，包括瓦楞柱和石墙常年裸露。从水边广阔的平台步入酒店，历史的建筑材料与这里时尚、鲜艳色彩的家具和装置并列，吸引年轻的或有颗年轻的心的客人。J+W事务所特意为该项目创作了大型的裸体画装饰室内墙壁。吸引人注意力的墙面处理和其他一些稍微有些暴露的设计元素将酒店重新定位为迷人的、令人向往的现代富人的度假村。

步入客房，设计大胆且不复杂：运用了红色块，经典的电影明星剧照印在棉面料上作为床头上部的背景墙。当地的石材被用在墙壁上，并同时采用了米色的砂石和石板做地面材料。室内的设施只用了黑色和红色两种色调。浴室的手盆也是红色的。卧室和浴室间的玻璃墙给人以空间扩大之感。没有门的间隔，只是在玻璃上有几条半透明的磨砂带，给人一些私密感。

除了45间客房和9间套房外，J+W事务所还为酒店的新酒吧设计了背光照明的绿色蚀刻玻璃立面和柚木台面的吧台。接待台的另一端是一个一体化的餐厅，它和一个扩大的平台相连，平台也是餐厅的室外就餐区。

Location
Hvar, Croatia
DESIGNERS
Jestico + Whiles
COMPLETION
2008
PHOTOGRAPHERS
Jestico + Whiles

项目地点
克罗地亚.赫瓦尔
设计师
Jestico + Whiles
完工时间
2008
摄影师
Jestico + Whiles

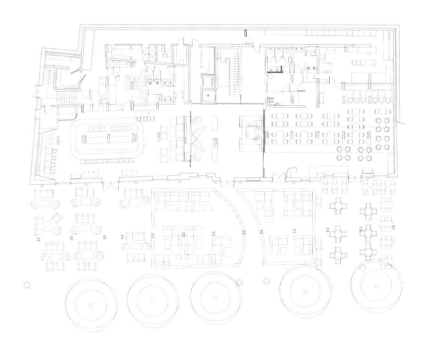

Haymarket Hotel

海玛特酒店

The Haymarket Hotel is situated off London's Haymarket in the heart of the theatre district and right next to the Haymarket Theatre Royal and just steps from Trafalgar Square, the National Gallery and St. James's Park.

A bold step away from cookie-cutter minimalism, Haymarket Hotel fuses contemporary and classical references in an ultra-central London location. A landmark building designed by the legendary John Nash, the building comprises 51 bedrooms and suites plus one exclusive townhouse property.

The façade features a dramatic row of columns that run the length of Suffolk Place. Interiors are a remarkable combination that honours the building's noble lineage while updating it with co-owner Kit Kemp's "modern English" interpretation of interior design. The lobby is a clean airy space featuring a large stainless steel sculpture by Tony Cragg and paintings by John Virtue, while the downstairs pool features a spectacular lighting scheme that uses strong contrasts of light and shadow. Guestrooms are individually furnished and feature custom-made pieces, avoiding the "designer formula" look and emphasising rich texture and colour. The two private townhouses have direct access to the hotel and all its services and facilities, offering large-scale luxury living in a hotel context.

海玛特酒店位于伦敦海玛特剧院区的中心地带，紧临海玛特皇家剧院，与特拉法加广场、国家艺术馆和圣詹姆士公园仅几步之遥。

海玛特酒店在设计上大胆走出最简单艺术派主义的固定模式，强调在伦敦超中心的位置上的现代和古典两种特色。这个地标性的建筑是由带有传奇色彩的约翰·纳什设计的，这幢大楼包括51间客房和城市住宅。

酒店正门矗立一排引人注目的圆柱，一直到萨福克宫。室内装修一边按照基特·肯普自己的理解表现现代性，一边保留大楼贵族血统的气质，两者惊人地结合起来。酒店大堂是一处通风的简洁空间，装饰着一个托尼·克拉格的不锈钢雕像和约翰·沃德的绘画作品。拾级而下，酒店的游泳池采光明亮，光亮和阴影的对比强烈。客房的布置按客户要求进行，带有个性化的特点，这样就能避免千篇一律的"设计师公式"产生的外观效果，同时强调丰富的质感和色彩。两座私密的城市住宅可以直接进到酒店，所有内部的服务和设施都可以提供酒店豪华的居住方式。

Location
London, UK
DESIGNERS
Kit Kemp
COMPLETION
2007
PHOTOGRAPHERS
Kit Kemp

项目地点
英国,伦敦
设计师
基特·肯普
完工时间
2007
摄影师
基特·肯普

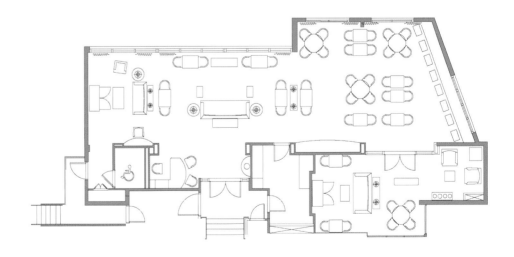

Hotel Skt Petri

第一佩奇酒店

Situated on a beautiful street in Copenhagen's quaint, trendy Latin Quarter, the First Hotel Skt Petri is a fine example of superior minimalist Scandinavian design-with a warm welcoming glow. Downtown Copenhagen, is next to the pedestrians' streets and City Hall Square. Nørreport Metro and S-train Station is only 5 minutes walk from the hotel connecting you to Copenhagen International Airport in just 12 minutes. It's a renovated 1930s' department store named after the famous church nearby with 268 rooms including 27 suites.

The mezzanine lobby is filled with clean lines and curves, and the glamorous large rings of light hanging overhead crown the soaring space. The Brasserie Petri lobby restaurant is open in summer to a soothing courtyard garden; the lobby also showcases the elegant nightspot Bar Rouge, one of Copenhagen's most exclusive lounges.

One of Denmark's leading visual artists, Per Arnoldi, selected a tricolour scheme for both public and private spaces: bright whites contrasting with his signature cool blues and vivid reds adorn the hotel's interiors down to the smallest accessory. This kind of attention to detail is also evident in impeccable, personal service that makes the 268-room hotel feel much smaller.

第一佩奇酒店坐落在哥本哈根古怪时髦的拉丁区，一个美丽的街道上，是简约的斯堪的纳维亚设计风格的杰出代表，带着一种暖意吸引客人。酒店位于哥本哈根市区，毗邻步行街和市政厅广场。从酒店到那拉堡特地铁站和S火车站步行仅需5分钟，去哥本哈根国际机场只需12分钟。该酒店原来是一家因附近教堂得名的百货商店，在20世纪30年代重新翻修。酒店的268间客房包括27个套房。

带有夹层楼面的酒店大堂充满了整齐的直线和曲线，头顶悬挂的迷人的巨大灯环增加了整个空间的贵族气息。百事丽皮特尼大堂餐厅，夏季的时候一直开到使人感到舒心的乡村花园里。大堂也展现出夜总会胭脂酒吧优雅的休息大厅，这个休息大厅是哥本哈根最有特色的。

丹麦最重要的视觉艺术家之一佩尔·阿诺迪为酒店的公共区域和私人空间选择了一种三色方案：用明亮的白色与他代表性的酷蓝色形成对照以及生动的红色装饰酒店的内部，小到最小的配件。十分明显，这种对细节的注意也运用到挑不出任何瑕疵的服务上，酒店的268间客房显得要小得多。

Location
Copenhagen, Denmark
DESIGNERS
Erik Møllers, Tegnestue Millimeter Arkitekter, Per Arnoldi
COMPLETION
2007
PHOTOGRAPHERS
Erik Møllers

项目地点
丹麦,哥本哈根
设计师
埃里克·莫勒,泰格尼斯特米利米特建筑事务所,佩尔·阿诺迪
完工时间
2007
摄影师
埃里克·莫勒

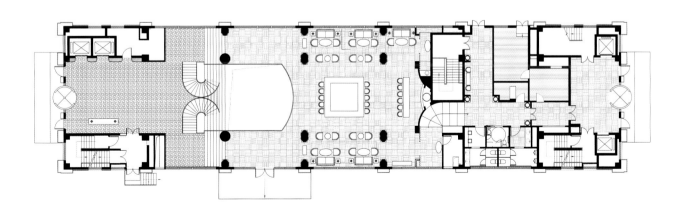

The Grand Daddy Interesting

老爸趣味酒店

This townhouse was constructed in 1870, acquired a Georgian façade in 1905, and subsequently went through a bewildering, yet brilliant, series of reworkings. In preparation for its current role as Cape Town's The Grand Daddy (formerly the Metropole), François du Plessis made the most of the resulting funky mixture of retro, modern and classic, adding a vivacious touch of African chic.

The lobby is served by South Africa's oldest working elevator, and dominated by the passion and decadence of the red colour. International guests can settle into the ground floor M Bar & Lounge's plush, poppy hues, or be transported to the 1960s in the street level M Café, in chocolate and lime green. Even in the clean-lined Veranda restaurant with its absorbing view of Long Street, bright colour juxtapositions remain the rule. The 25 guest rooms give their visitors the contrast of peaceful luxury brought up to the present day with a soothing, textured minimalism and high-tech appointments. The building might have all the character of maturity and experience, but the hotel that inhabits it boasts an undeniably youthful energy that sparkles exuberantly throughout its spacious chambers and broad corridors.

这个排房住宅于1870年建成，1905年其外观弄成乔治亚建筑风格，后来又经历了一系列的虽令人费解却又辉煌的重修重建，现在它又成了开普顿的老爸酒店（旧称京华国际）。为此，设计师弗朗克伊斯·杜·普莱西斯充分地把复古、现代和古典这些风格巧妙地组合在一起，为建筑物添加了一丝活泼的非洲高雅色彩。

酒店大堂有南美最古老的电梯，主打代表激情和颓废的红色。国际宾客可以被带到一楼的M酒吧和休息室，一片豪华的、芙蓉红，也可以被带到20世纪60年代M咖啡厅，全是巧克力和石灰绿色。甚至在可以观赏到长街美景的线条简洁的阳台餐厅也是奉行各种明亮颜色并置的原则。酒店内设25间客房，使参观者体验到把宁静的奢华带到现在生活让人放松和高科技运用的对比。整个大楼既具有所有成熟的特点，同时也有年轻活泼的气质，活力通过宽敞的房间和宽阔的走廊散发出来。

Location
Cape Town, South Africa
DESIGNERS
Tracy Lynch, Robin Sprong
COMPLETION
2008
PHOTOGRAPHERS
Tracy Lynch, Robin Sprong

项目地点
南非,开普敦
设计师
特蕾西·林奇,罗宾·斯布朗
完工时间
2008
摄影师
特蕾西·林奇,罗宾·斯布朗

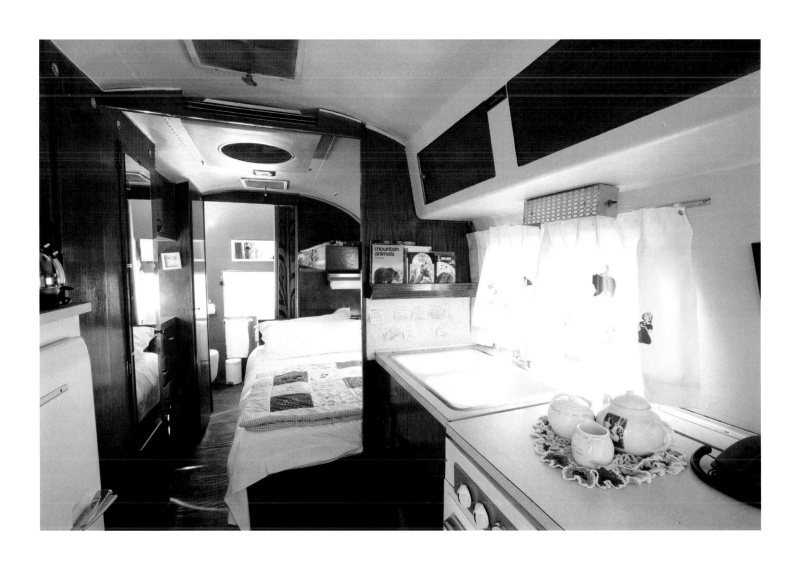

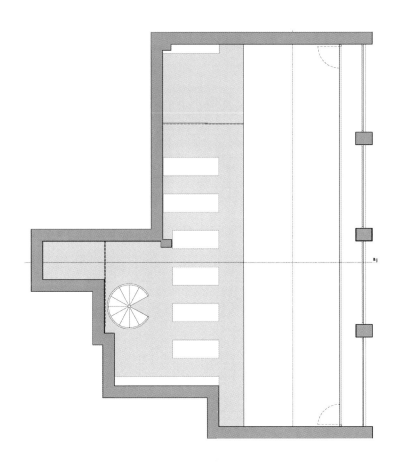

Marriott Marquis

万豪伯爵酒店

The $138 million renovation project of this Atlanta landmark hotel respects the original Portman design while enhancing the existing features and relocating all food and beverage outlets off the lobby bar to create a hub of activity.

Phase I of the renovation was completed in August 2007 and included the opening of four new food and beverage facilities: 'Sear' fine dining restaurant, 'High Velocity' sports bar, 'Pulse' lobby bar and a new Pool Bar. The 'Pulse' lobby bar is the focal point and brings new life into the hub of the hotel. The fifty-foot structure is clad in resin panels that are backlit and change colour throughout the day to reflect the 'pulse' of the property.

Phase II of the renovation was concluded in August 2008 and centres on the addition of the new Atrium Ballroom, the renovation of the existing Marquis Ballroom, the renovation of the main lobby and front desk, and the brand new fitness centre and spa. The main lobby received a completely new front desk also curved to reflect the architecture of the atrium and designed in sections to allow employees to easily reach the guests.

耗资13.8亿美元的改造亚特兰大地标性建筑——万豪伯爵酒店项目小组尊重原来波特曼的设计，并提升现有的酒店特色，重新布置所有食品和饮料的卖场，远离大堂酒吧处，让大堂成为活动的中心。

一期的改造于2007年8月完成，四个新的食品和饮料卖场设施开始营业：美式"煎炸"餐厅、"高速运动"酒吧、"脉动"大堂酒吧和台球酒吧。"脉动"大堂酒吧是项目的重点，它将新的活力带入酒店核心部分。15.24米高的结构用树脂板建成，利用背光照明，让它在一天中变换色彩来反映酒店的"脉动"。

二期的改造于2008年8月完成，加建了一个新的中庭舞厅，改造了现有的酒店舞厅、主体大堂和前台、品牌健身中心和桑拿房。主体大堂处新建了一个前台，也是采用曲面来反映中庭的结构，并分块设计，以便员工近距离地接待客人。

Location
Atlanta, USA
DESIGNERS
Thompson, Ventulett, Stainback & Associates
COMPLETION
2008
PHOTOGRAPHERS
Thompson, Ventulett, Stainback & Associates

项目地点
美国,亚特兰大
设计师
Thompson,Ventulett,Stainback & Associates
完工时间
2008
摄影师
Thompson, Ventulett, Stainback & Associates

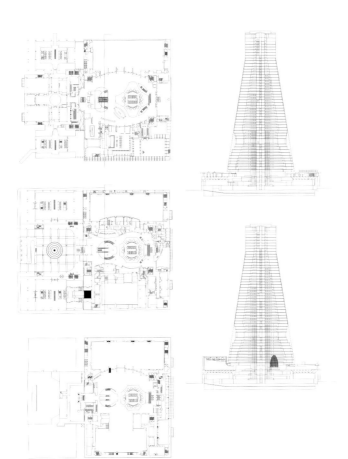

Peter Restaurant

彼得餐厅

Occupying the entire 24th floor of The Peninsula Tokyo, the restaurant features sweeping 360-degree views of Tokyo and the famed Imperial Gardens. The designers incorporate the visual stimulation of the splendid view into Peter's design with specially commissioned artwork to enhance the scene. Curved windows surrounding the walls are interspersed with special acrylic panels by artist Marc Littlejohn, featuring embedded mirror strips in an abstract grid pattern which capture the cityscape lights. During the day, elegant panels of eggplant and gold fabric gracefully filter the light.

A vestibule entrance features sculpted walls of folded black metal slabs pitched at exaggerated angles. Crafted by D'Art, this dramatic installation is lit from the top illuminating subtle infinity patterns scored onto the surface.

Integrating technology and motion graphics, the restaurant's creative design includes an interactive video wall with projected images of Hong Kong behind the raised platform. The elevated "stage" features two semi-private dining areas which are partially concealed by curved artistic screens to veil guests.

In the banquet room, an expansive artwork by Hirotoshi Sawada features a suite of silver metal strips. To create a nest shape looking, the sculpture covers the whole ceiling and flows to parts of the wall.

高居于东京半岛酒店的24层楼上。餐厅四面都可以看到东京的景色和著名的皇居。设计师融合了壮观的外部视觉景观，将内部用特别定做的艺术品装点。环绕墙面的曲面窗是用艺术家马克·小约翰设计的丙烯塑料板点缀的，镶嵌上去的镜面条以一种抽象的格栅样式布局，捕捉城市的风景线。白天，光线穿过紫色、雅致的嵌板和金色的织物，呈现唯美的效果。

入口门庭的雕塑墙用折叠的黑色金属条镶嵌，搭成夸张的角度。从上部照明，D'Art精心设计的浪漫墙面显现了一种精妙无限的折叠图案。

在抬升的平台就餐区后面是一面交互式电影墙，投影仪向客人展示香港的景色，这是餐厅里的一处创新设计。抬升的"舞台"处有两个半私密就餐区，用曲面的艺术屏做遮挡视线的围屏。

在宴会厅里，一件大幅的、由日本著名艺术家Hirotoshi Sawada制作的艺术作品——用银色金属条创造了一个鸟巢的空间形状，艺术品覆盖了整个天棚和部分墙面。

Location
Tokyo, Japan
DESIGNERS
Yabu Pushelberg
COMPLETION
2007
PHOTOGRAPHERS
Yabu Pushelberg

项目地点
日本.东京
设计师
雅布&普歇尔伯格公司
完工时间
2007
摄影师
雅布&普歇尔伯格公司

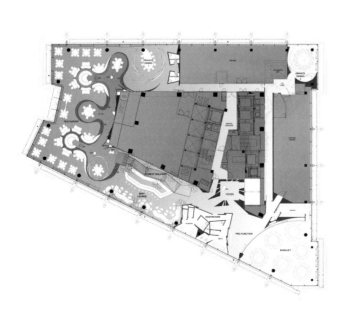

Barolo Ristorante

巴罗洛餐厅

The client requested a restaurant that blends old world charm and modernism in an extremely narrow 2,300sq.ft space with very high ceilings. The designers created a variety of experiences within the space, designing a large communal table as the central focus and lowering ceilings to create spatial hierarchy. Strategically placing expansive mirrors provided the illusion of a much larger space.

Lighting design played a major role – ornate crystal chandeliers light the communal table while minimal spots illuminate the massive wine cases, creating a heightened sense of drama. Backlit onyx bar counters and plinths provide a soft, subtle glow.

The budget and timeframe were very aggressive, so the designers specified readily available simple furnishings and materials throughout and strategically select the "big ticket" items which become conversation pieces.

The mixture of diaphanous white sheers, drippy candles, baroque mirrors, green velvet booths and dark concrete floors created a feeling of opulence and style. The end result is a soulful space that does not upstage the food but provides a glorious dining experience while reflecting the owner's Italian heritage.

客户希望餐厅的设计能够将古典的美和现代主义结合在这个占地214平方米，顶棚高挑、异常狭窄的地方。设计师在这里创造了一系列的空间体验：将一个大的公共餐桌作为餐厅的中心；降低了顶棚的高度，以创建出空间的层次感；策略地布置了几面大镜子，如同空间被扩大了一般。

在这里，灯光的设计起了很大的作用：华丽的水晶枝型灯照亮了公共餐桌，而小的灯光则为大面积的葡萄酒柜提供照明，抬升了一种戏剧化的氛围。玛瑙红吧台和柱基处则采用了平缓、精妙的背光照明。

项目的经费和时间都非常有限，所以我们为酒店定的都是简单的、现成的家具和材料，并策略性地选择了几件"大牌"的物品，做为人们关注的焦点。

透明的白纱帘、流泪的蜡烛、巴洛克风格的镜子、绿色天鹅绒的隔间、黑色的混凝土地面，这些给客人一种庄严辉煌而又时尚的感觉。餐厅风情卓绝的空间不仅提升了食物的品味，而且提供了一个美好的就餐体验，同时也体现了餐厅主人传统的意大利风范。

Location
Seattle, USA
DESIGNERS
Corso Staicoff, Inc
COMPLETION
2006
PHOTOGRAPHERS
Sally Painter, Thomas M. Barwick

项目地点
美国，西雅图
设计师
科索斯泰科夫股份有限公司
完工时间
2006
摄影师
莎丽·蓬特，托马斯·巴维克

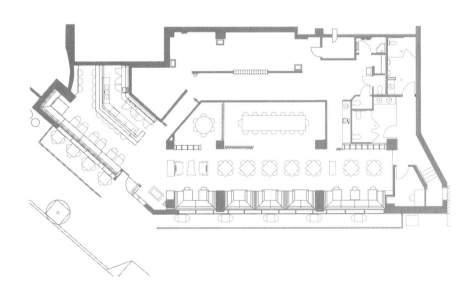

Sushi Restaurant

寿司餐厅

Sushi Restaurant located in a Shopping mall with plenty of natural light and a "T" floor plan.

The Sushi bar was designed to function as a main entrance to the restaurant. A mezzanine floor was located above the Sushi Bar given a lower ceiling to this area. Arthur Casas designed a backlit ceiling at the Sushi bar area to estabilish a relationship with the Japanese culture and the cuisine. At the same time the backlit ceiling signalised the main entrance catching the public attention and amplifying the lower ceiling space.

The dining area has a higher ceiling- approximatley 6 metres or 19.6 feet and a big window giving plenty of natural light to the space. The dining table has received Tom Dixon's pendant lights and directional lights were placed along the walls, washing the artwork and the plaster finished walls creating a more dramatic and theatrical effect to the dining area.

寿司餐厅位于一家购物中心内，自然光线充足，T型建筑平面图。

寿司吧的设计要表现出其餐厅主入口的功能。在寿司顶部安上一个夹层楼面，使这个地方的天花板要低一些。亚瑟·加萨斯在寿司吧区设计了一个从背后照亮的天花板，使日本文化和餐厅料理联系起来。同时这种从背后照亮的天花板也突出了主入口的位置，吸引人们的注意，增大了低天花板的空间感。

就餐区的天花板要高出一些，大约6米或19.6英寸高。一块大玻璃窗使自然光线充分照射到室内。公共餐桌上有汤姆·迪克森的吊灯，沿墙还放有一些笔直的灯具照射艺术品和石膏板的墙壁，给整个就餐区增加了戏剧性的效果。

Location
Sao Paulo, Brazil
DESIGNERS
Studio Arthur Casas
COMPLETION
2008
PHOTOGRAPHERS
Studio Arthur Casas

项目地点
巴西,圣保罗
设计师
亚瑟加萨斯设计工作室
完工时间
2008
摄影师
亚瑟加萨斯设计工作室

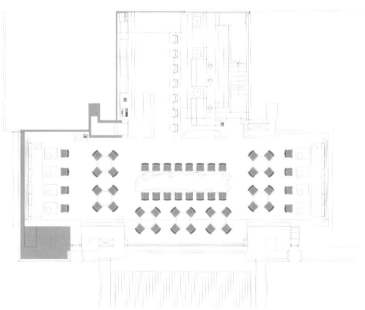

Restaurant Mangold

曼古德餐厅

The restaurant in located in an 1870s' building which is part of the four-star hotel "Gastwerk" in Hamburg's Altona. The redesign of the restaurant is part of a relaunch project for the 10th anniversary of the hotel in january 2010.

Basically the design had to be integrated in the old existing brick architecture. So the materials need to be classial. The most important thing was the haptic feeling of whatever the guest touches, green velvet, natural tanned leather, brass and oyster coloured massive oak with a somewhat rough look and feeling. Partitions with an industrial design seem to have always been there.

A big bar which shelters the breakfast buffet in the morning changes into a frontcooking desk in the evening, where the guests can order his fish directly and talks with the cook over the way of preparation.

The restaurant is equipped with generous and very comfortable benches. The seating, consistently with armrests, doesn´t simplify the choice of where to sit.

In the centre stands a classical table with a maple desk top where all-time the newest journals and magazines are displayed.

The table is surrounded by four columns in form of a carillon.

餐厅位于一座1870年建起的大楼里，该大楼是汉堡亚托纳区煤气厂酒店的一部分。对餐厅的重新设计是为庆祝2010年该酒店成立10周年举行的庆典活动的一部分。

该设计基本上要与原来砖质建筑物结合一体，所以选择的材料必须是古典的。最重要的一点是酒店客人无论触摸到什么东西，东西必须具有触感。绿色的丝绒、天然的硝制皮革、黄铜和外表粗糙的牡蛎色的高大的橡树。工业设计的隔墙表面的玻璃似乎一直就在那里，而不是现在才设计出来的。

早晨供应自助早餐的一间大酒吧晚上就变成了一个现场烹饪厨台，酒店客人可以直接点鱼吃，还可以和厨师聊聊备料方法。

餐厅配有非常宽大舒适的长椅，座位上的扶手并没有简化挑选座位的过程。

在中心有一张古典餐桌，枫树制成的餐桌表面上总是会放一些最新的期刊和杂志。

餐桌周围是钟琴造型的四个圆柱。

Location
Hamburg, Germany
DESIGNERS
3meta design devision
COMPLETION
2008
PHOTOGRAPHERS
Andreas Brücklmair

项目地点
德国.汉堡
设计师
3元视觉设计工作室
完工时间
2008
摄影师
安德里亚斯·布鲁克梅尔

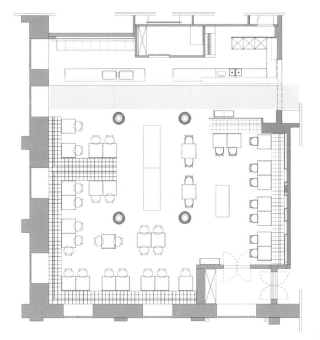

Pan-American Restaurant

全美洲餐厅

Cordua has engaged the architectural firm of Studio Gaia, led by its founder Ilan Waisbrod, to design the modern Pan-American Restaurant. Waisbrod & company are responsible for such high-profile projects as the W hotels in Seoul and Mexico City, Tao Las Vegas restaurant in the Venetian, the 40/40 Club owned by Jay-Z, and Wave restaurant in the W Chicago hotel. Their style is crisp and (mostly) minimal, with strong linear elements and provocative seating, often including chairs and ottomans in organic egg, pod or hemispherical shapes.

Michael Cordúa teamed up with Studio Gaia, a renowned New York City architectural firm, to create sleek interiors and bold, imaginative elements that complement the romantically provocative menu.

Descend the stairway into the dining room where vivid reds and oranges are woven together with natural accents of stone and wood, overlooking The Woodlands Waterway. From the hand crafted wood tables to the slow tumble of water down three water walls to the oversized banana leaf murals overhead, guests are transported to the tranquility of a tropical rainforest.

科杜阿委托宜兰·威斯布鲁德成立的盖亚建筑工作室设计一个现代的全美洲的餐厅。威斯布鲁德的公司负责承办过像首尔、墨西哥城的W连锁酒店、威尼斯的陶拉斯维加斯餐厅、杰伊开办的40/40俱乐部和在芝加哥的W酒店里的波浪餐厅等知名度高的项目。他们的风格清新、简单、抽象，大部分都是采用色彩强烈的线条和经常包括蛋形的、近圆柱形的或半球面状的椅子和脚垫在内的刺激感官的椅饰，来表现设计风格。

盖亚工作室是纽约一家著名的建筑公司，迈克尔·科杜阿与盖亚工作室合作，一起设计出时髦的室内装饰，运用大胆富有想象力的元素来创造出浪漫的吸引人的菜单。

沿着楼梯下到餐厅，鲜艳的红色和橘红色与自然突出的石头和树木交织在一起，俯瞰林地水路。从手工制成的木头桌子，到顺着三座水墙缓慢流下的水流，再到头顶上巨大的香蕉叶壁画，顾客被引到一片热带森林的宁静安谧之中。

Location
New York, USA
DESIGNERS
Studio Gaia
COMPLETION
2007
PHOTOGRAPHERS
Courtecy of Studio Gaia

项目地点
美国,纽约
设计师
盖亚工作室
完工时间
2007
摄影师
盖亚工作室

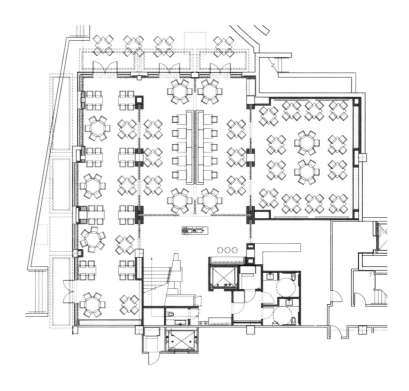

Fremantle Media

弗里曼特尔传媒办公大楼

With the proposal of zen Dens, water features, organic vegetable gardens, baachi sand pits, barbeque entertaining and yoga meditation huts, it was Design Director Andrew Cliffe of The World is Round who won the right to create a playful media space for their combined home.

The first floor houses the community café allowing direct access to the northern early morning sun filled balcony. Located adjacent to this is the client focused media boardroom. It's used as a device for moving visitors through the building, but it also seeks to disarm the visitor by relaxing them with coffee and flipping the old psychology of hiding the guts of organisation right on its head, as the work place is in open view to the visitor.

Slick finishes such as Axoltle, quartz carpet and the upholstered Moov™ wall system add playful elements of texture and movement to the environment, while the unique feel on each floor and unexpected meeting rooms maintains the confrontational style Cliffe is known for.

"The space is functional, playful, human, welcoming and glossy."

该项目的设计包括水系、有机植物园、沙坑、烧烤娱乐和瑜伽冥想茅屋等。世界是圆形公司的设计总监安德鲁·克利夫获得了为他们共同的公司新家创造这个充满动感的媒体办公大楼的权利。

二楼有公用咖啡厅，直接通往一大清早就充满阳光的阳台，接下来的是客户的媒体会议室。这样安排的目的是来访者可以由此进入大楼，也是用咖啡来使来访者放松。由于敞开办公，使来访者消除公司隐藏实力的旧的观念。

圆滑的石英地毯和加软垫的摩伊墙壁设计增加了环境的质地和动感，每层地板和出乎意料的会议室保持了克利夫一向出名的对抗风格。

"整个空间具有功能性、娱乐性和人性，吸引人，也具有光洁性。"

Location
New South Wales, Australia
DESIGNERS
TWIR
COMPLETION
2007
PHOTOGRAPHERS
The World is round Pty Ltd

项目地点
澳大利亚,新南威尔士
设计师
TWIR设计工作室
完工时间
2007
摄影师
世界是圆形产权有限公司

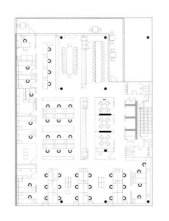

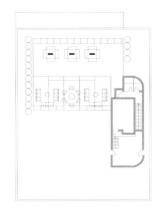

Papsa Showroom

帕帕莎陈列室

Papsa, distributor of Haworth in Mexico, opted for reshaping its offices carrying out a significant change in various aspects.

The space should comply with two functions at the same time: it should support the operation of the company and function as a furniture showroom.

Water, marble and bamboo are three elements that give the visitor a fresh welcome, inviting him to the first meeting point in the tour: the "war room". To raise awareness and promote responsibility towards our environment and Mexico, the user receives in this space information on sustainability, the development of this project and the LEED-CI certification process.

The materials used in the project are sustainable-green, ranging from the false floor to the furniture, including: carpet, sheetrock, MDF, paint, sealers, adhesives, wheat finish work area covers (quickly renewable), among many others. One of the innovations in this project is an unconventional method for putting carpets into place (tactile).

With an aim at improving the indoor air quality and providing maximum flexibility to the showroom, a high false floor was used to allow injection of air conditioning through it, making energy consumption more efficient, and providing greater comfort to users.

帕帕莎是墨西哥霍沃思的经销商，他们在多种方案中选择重新改装办公室，在各方面都具有重要的意义。

空间建筑事务所应该遵从以下两种功能：应该支持公司的运作，并且发挥家具陈列室的功能。

水、理石和竹子这三种要素给人一种清新的感觉，引人来到旅行的第一站："作战室"。参观者可以获得可持续性相关的知识、了解项目的进展情况和获得商业建筑室内能源和环境设计的领袖认证的过程，从而加强对环境和墨西哥的责任感。

项目使用的材料都是绿色环保的，从夹层地板到家具，包括地毯、石膏板、主配线板、涂料、胶、黏合剂、工作区完成后覆盖的小麦（可快速更换）等等。该项目的创新之一就是放置地毯的非传统的方法（有触感）。

为了改善内部空气质量，给予陈列室最大的灵活性，使用了夹层地板。这样空调可以吹入，使能源消耗更为有效，提高空气质量，提供更大的舒适感。

Location
Mexico city, Mexico
DESIGNERS
Space
COMPLETION
2007
PHOTOGRAPHERS
Santiago Barreiro

项目地点
墨西哥 墨西哥城
设计师
空间建筑事务所
完工时间
2007
摄影师
圣地亚哥·巴雷鲁

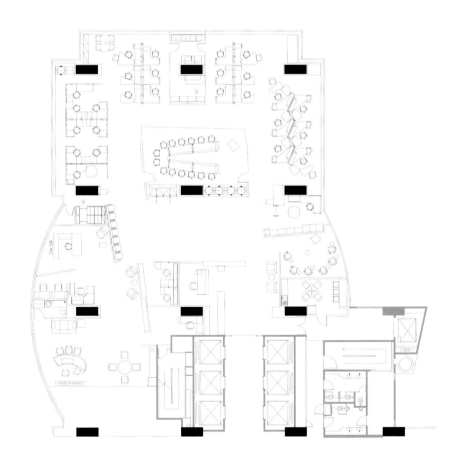

Ecomplexx Officez

Ecom公司的综合办公大楼

Terms and concepts from Ecom's network, e.g. "Plug in" or "Backbone" are to play an important role in the company's future architecture to support and develop its own identity. No changes have been made along the attractive windows to retain the impressive generosity of the "loft feeling". The corridor and operating space runs along a façade which is suffused with light. The concept of the central "Backbone" tries to focus all changes on this connecting core area which has the potential for a lot of different uses. Many workstations are connected to the "Backbone", which creates an open and communicative office structure.

The entire expanse of the loft's historic window façade can now come fully into effect while the multifunctional Backbone offers all the essentials, e.g. stacking, archive, coat room, features for relaxation like seating areas, the bar as well as the entire technical installation. The Backbone was built in the form of serial shelf-like furniture, as a mega structure of untreated chipboard to comply with the extremely low budget. Special material has been used for individual shelves to ensure pleasant acoustics, and soft seating areas have been built as well as lounge-like meeting areas surrounded by soft curtains. Glass walls are completing noise protection. The detached Box (Sanitary Rooms, Reception, Kitchen, Meetingspace) is coated with dark paint and is used as a huge blackboard for making notes.

The 38 workstations form an innovative and communicative office structure with a good social atmosphere for a low cost. The dialogue between the historic material and the new additions using lightweight design and considering the geographical separations, combine the demands of monumental protection with the demands of modern rooms for innovative companies.

Ecom公司的术语Plug in 和Backbone在这个公司的未来建筑中起到了十分重要的作用，它们用来支持和发展其自身的特点。为了保留阁楼出众的宽敞感，设计师并没有对这些吸引人的窗户进行改动。走廊和操作空间被安排在了光线充足的正面。中央脊骨的设计理念强调这个联通的核心区域的所有变化。这个联通区域可以有许多潜在的不同用途。许多的工作室都与这个脊骨相连，从而创造了一个开放的、联系的办公结构。

阁楼上历史悠久的玻璃墙面得以被充分利用，而多功能的脊骨结构则提供了所有的必须设备，如仓库、档案室、更衣室，放松设备如休息区、酒吧以及整个技术设备。这个脊骨结构是一系列的架式的家居，这是为了使这个巨大的由天然硬纸板组成的结构能够符合这个项目极低的预算。设计师在每个架子上使用了特殊的材料已达到最好的声效。这里同时还设计了舒适的休息区以及一个由窗帘围起来的像等候室一样的会议区。玻璃墙面用来阻挡噪音。"独立的盒子"（卫生间、接待室、厨房、会议空间）被涂上了暗色调的油漆，它就像一个巨大的供人们记笔记的黑板。

38个工作室构成了一个富有创意的、联系紧密的办公结构，这个建筑成本很低的结构提供了一个良好的社交氛围。新增的结构采用了轻质材料，并考虑了地理的分隔，它与保留下来的具有悠久历史的材料的对话使保护历史遗产的需求与建造现代创新公司的需求结合在了一起。

Location
Wels, Austria
DESIGNERS
Xarchitekten
COMPLETION
2009
PHOTOGRAPHERS
Max Nirnberger

项目地点
澳大利亚,维尔斯
设计师
Xarchitekten建筑事务所
完工时间
2009
摄影师
麦克斯·尼恩伯格

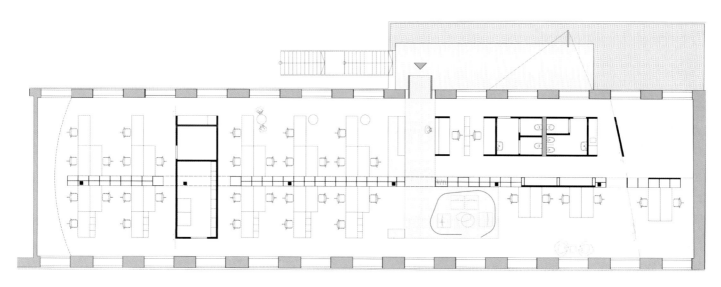

Toyota Tsusho

丰田通商

The office interior of Toyota Tsusho, one of biggest trading company in Japan. Sinato mainly planned the visitor space.

The location of this office is on the 33rd floor of a skyscraper in Nagoya city and the plan which utilises a fine view of the city is required. But the space had to be filled with continuation of meeting room and it tended to make the space closed. So the challenge was to make the space opened in this condition.

The designers made a passage of eyes which goes diagonally through the visitor space. People can feel the whole space in this long passage and they can feel the space extends to the outside beyond a long distance of eyes.

As for the wall the designers made gradation from opaque white to half mirrored glass. They also made gradation of colour for the floor. These are the promotion of the passage, but at the same time they blur the line of passage like a line of sunlight.

这是日本丰田通商公司的办公内部装修，丰田通商是日本最大的一家贸易公司。设计师主要设计规划了客户来访区。

办公地点位于名古屋一幢摩天大楼里的第33层，因此要求设计师的规划利用到城市的美景。但是空间内必须设有很多会议室，因此要求一定的封闭性。因此我们的挑战是在这种条件的限制下保证空间的开阔。

设计师设计出一个视觉的通道，呈对角线地穿越整个客户来访区，这样人们可以有一种这个空间在长长通道的很远处延伸到室外的感觉。

关于墙壁，设计师从不透明的象牙白色渐渐过渡到磨砂玻璃，在地板的颜色选择上也采用过渡渐近的方法。整个设计既有通道的提升，同时也把通道设计成像一缕阳光那样模糊其界线。

Location
Nagoya, Japan
DESIGNERS
Sinato
COMPLETION
2008
PHOTOGRAPHERS
Takumi Ota

项目地点
日本,名古屋
设计师
Sinato公司
完工时间
2008
摄影师
匠大田

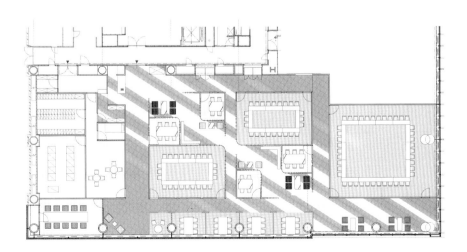

103

Wieden+ Kennedy

文登+肯尼迪公司办公大楼

The company requested a building addition to house the corporate office. The site is the leftover space between the existing sales office building and the existing plant building.

The existing office building and existing plant each read as distinct volumes with no relationship. The addition was designed to be a third piece to unify the building, constructed from materials from both the existing office building and the existing manufacturing plant.

The solution evolved into a large open space adjacent to a curtain of glass with private offices, back of house rooms along the perimeter, and work spaces on the mezzanine above. All of the interior rooms look out to the large open space through large expanses of glass. The circulation is through the large open space with a grand high-tech staircase that connects to the mezzanine. The solution clarified the departments and demarcated spaces, while simultaneously opening the spaces to each other and weaving spaces together. Natural light was brought into the interior through two large slot openings in the roof to allow maximum amount of light to enter the space. The skylights have been designed with movable reflectors that screen and reflect light to all spaces in the office.

公司要求增加一座大楼作为公司的办公楼，项目地点是现有销售大楼和工厂大楼中间的剩余空间。

现有办公大楼和工厂大楼彼此独立，因此设计出另外一块大楼把它们统一起来，建筑材料取自现有办公和工厂大楼。

解决方案最终变成一个大的开阔空间与一个玻璃幕墙相邻，私人办公室在屋子的最后，工作区在上面的夹层楼面上。所有的室内房间都可以通过大片的玻璃看到外面的空地。空气通过与夹层楼面相连、带一个大楼梯的宽阔的空间流通。这样就把部门和划分界限空间区别开来，同时也使各个空间彼此相通，互相开放。大量的自然光通过屋顶两个狭缝开口照进来，把天窗设计成了可移动的光线反射器，把光线反射到办公室的每个空间。

Location
London, UK
DESIGNERS
Jump Studios
COMPLETION
2008
PHOTOGRAPHERS
Gareth Gardner

项目地点
英国,伦敦
设计师
跳跃设计工作室
完工时间
2008
摄影师
加雷斯·加德纳

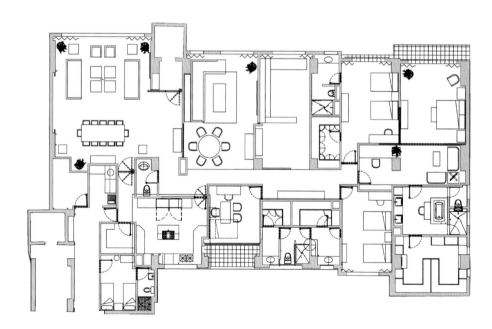

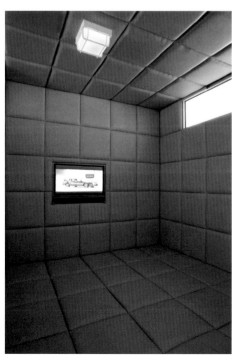

980 5th Avenue

第五大道980号

The clients are an empty-nest couple, with a big home in the suburbs and a self-declared 'tchatchki' addiction. This apartment is meant to function as an escape from their ordinary life that is filled with collections and objects.

A gut renovation of the existing 2,500sq. ft apartment clears out the central space to visually link the dining room and the living room into one large space. The surfaces of the apartment are kept visually calm, their detail having been evacuated and aggregated into the Molding Wall. Small pockets of program occupy the vertical and horizontal surfaces of the remaining walls. These functional areas become the only articulation and decoration of the space. A vertical buffet serves the dining room, but when not deployed it creates a seemingly enigmatic pattern on the wall. The spectacular view to the Central Park is met with a tufted window seat that resonates the visual beat of the Molding Wall. Finally the client's desire for classical panelling is satisfied by muted panels.

项目的客户是一对"空巢"夫妇，在郊区有一座大房子，声称特别喜欢聊天。该公寓的设计要表达出从日常生活解脱的功能，他们的日常生活里充满了收藏的作品和物品。

该设计大胆地把现有的2500平方英尺的公寓的中心位置清空，这样餐厅和客厅在视觉上就被连成一个大的空间。公寓外观在视觉上具有宁静的性质，所有的细部特征都通过造型墙来表现出来。其他墙壁的外观无论是垂直的还是水平的，都显示出一些小的凹处。这些功能区成为整个空间唯一的联结点和装饰。餐厅里有一个垂直的餐台，但是在不展开拉出使用的时候，它看起来只是墙壁的一个神秘部分。中央公园的壮观景色和窗前簇绒椅子一起，与造型墙产生视觉上的共鸣。最后，使用色彩更为柔和的普通墙壁外框的面板，满足了客户的古典镶板的愿望。如果它们的对比强烈，可以通过涂白淡化。

Location
New York, USA
DESIGNERS
Christopher Stevens
COMPLETION
2008
PHOTOGRAPHERS
Michael Moran

项目地点
美国,纽约
设计师
克里斯托夫·斯蒂文斯
完工时间
2008
摄影师
麦克尔·莫伦

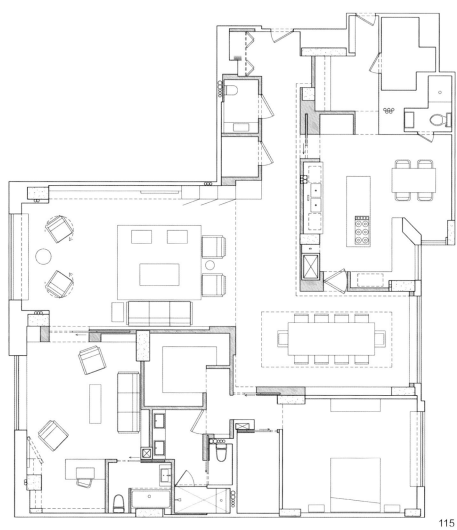

Tiburon House

蒂布伦住宅小屋

The centre of the house is an empty space, a courtyard-garden surrounded by a glass-walled gallery along its perimeter. The front door is on the western side of the building, at the point where the pavilions of the service areas and the guest quarters meet. A copper-clad door opens onto the long gallery that frames the view of the bay in the distance. This gallery is the real axis of the house, the linear element connecting the entrance and the beach. Along the gallery are sliding glass doors which open onto the south garden below.

Linearity is a recurring theme of the project. Besides the overall architectural layout, the linear development of the forms and space is repeated in the parallel lines of the wooden panelling on the exterior walls, the copper tubing sun-blinds in the gallery, the sliding panels of sheet aluminum, and the various objects and furnishings made to order for the project.

The custom-designed furnishings emphasise the use of copper. These include the big garage doors and the front door, some of the handles, the sun-blinds along the gallery, various lamps and an outdoor shower. All these elements were made by artisans in Italy. Copper is also used to cover all the roofs.

房子的中心是一块空地，有一处庭院花园，这个花园的周围由一个边缘是玻璃墙壁的画廊环绕。前门在房子的西部，也是服务区和客房相连处。一扇镀铜大门开到长廊处，把远处海湾的美景尽收进来。这个画廊是房子的真正的轴心，其线性结构把玄关和海滩连在一起。沿着这个画廊，一些滑拉玻璃门一直开到位于下面的房子南面的花园。

线形结构是这个项目中重复的主题。除了整个建筑布局以外，为项目定购的在外墙上面安置的木头嵌板的平行线、画廊里铜制管形材料的太阳薄膜、铝片的滑拉嵌板和房子的各种物体和家具陈设等等都体现出线形结构在建筑设计中的应用。

定制设计的家具陈设强调对铜的使用。这些包括车库大门和前门、一些把手、画廊的太阳薄膜、各种灯具以及屋外的淋浴器。这些东西都是由意大利的工匠制作出来的。设计中也用铜来遮盖所有的屋顶。

Location
San Francisco, USA
DESIGNERS
ANDREA PONSI architetto
COMPLETION
2008
PHOTOGRAPHERS
Richard Barnes

项目地点
美国,旧金山
设计师
安德烈·庞思建筑师事务所
完工时间
2008
摄影师
理查德·巴恩斯

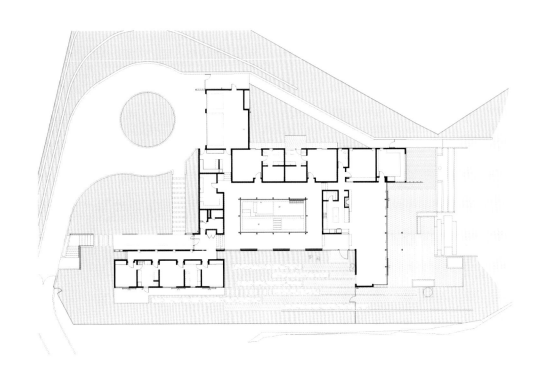

Private Residents House

私人住宅

This 550-square-meter Connecticut residence serves as the new home for a young family of six. Sited on two acres, the front of the house presents what seems to be a one and one-half storey façade. The rear of the house opens with generous walls of windows to the yard and nature preserve, creating an exhilarating relationship between indoors and outdoors.

The ground level is laid out in an open plan L-shaped configuration that embraces the large backyard with a terrace off the kitchen and a covered porch off the family room. The exterior living spaces blur the line between interior and exterior much as the large-scaled window openings merge indoors with the outside. Andres Cova worked with principal Eric Gartner to create the warm and muted interior palette that extends the natural tones of the landscape into the living areas, further linking the house with its environs. Hardwood floors flow continuously throughout the main level but custom silk and wool carpets ground the seating groups. Modern furnishings in inviting fabrics provide a synthesis between clean-lines and vibrant livability that the client desired, accommodating both aesthetic expectations and the active family and social life that define the household.

这个位于康涅狄格州的550平方米的住宅是一个六口年轻之家的新家。住宅占地面积2英亩，外观看起来像一座一层半的小楼。房屋后面的落地窗一直开到院子里，在屋内屋外创造出一种愉快的氛围。

平面图里一楼的设计取L形，厨房下台阶连着一个大的后院，卧室的下面有一个带篷的门廊。因为大玻璃窗的开阔把内部与外部连成一片，使外部居住空间模糊了室内与室外的界限。安德烈斯·科瓦与主设计师埃里克·加特纳一起创造出温暖的室内色调，把自然的风景引入室内，更加强了房屋与外部环境的联系。地上连续铺设实木地板，但是也有定制的丝绸和羊毛地毯。用这种织物设计的现代陈设把简洁线条和客户喜欢的充满生气的适于居住特点综合起来，既满足了对审美的要求，也反映出这个家庭积极的家庭和社交生活的特点。

Location
Stamford, USA
DESIGNERS
SPG Architects
COMPLETION
2007
PHOTOGRAPHERS
Frank Oudeman, John Gruen

项目地点
美国.斯坦福德市
设计师
SPG建筑事务所
完工时间
2007
摄影师
弗兰克·欧得曼,约翰·格龙

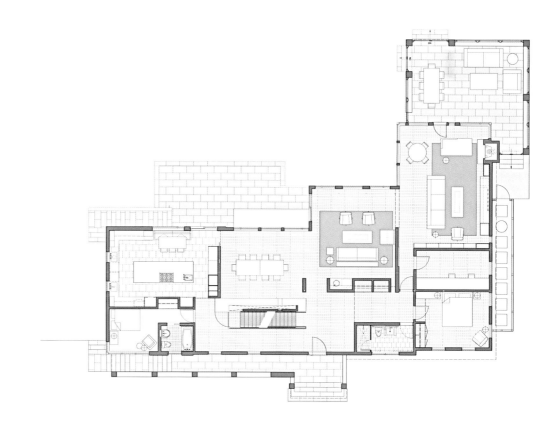

Vader House

维达住宅

The eastern and western façades of the extensions are encased in a shield of louvres. These peel back to reveal a folded internal environment of soft colours framed by exposed steel beams. Playfully splashes of deep red enliven the interior which is occasionally punctured by windows allowing a cinematic light to dance over the internal workings of the Vader House.

The refined material and colourful palette of the extension, wrapped in a heavy roof form distinguishes itself from the dark masonry clad terrace from which it emerges. These two opposing forms are united by a transparent glass corridor along the northern boundary wall, framing an outdoor courtyard.

Definition between these internal and external environments is barely distinguishable. Transparent bifold doors allow for constant physical and visual interaction, between these environs. The extensions is at once inside and out.

The open and seemingly simple nature of Vader House later reveals itself to be one of complexity and ambiguity.

The extension is created out of components that appear to have fallen at the eastern end of the site in a tetris-like manner. Unexpectedly a random tetris piece has lodged itself deep within the walls of the original building.

建筑延伸部分的东部和西部正面都有一个侧面带通风口的圆顶天窗的防护罩，后面是一个折叠的内部用外露钢梁框出来的色彩柔和的内部环境。深红色的斑点为室内增添了活泼的气氛，偶尔光线会透过窗户照射进来。

建筑延伸部分被一个厚重的屋顶结构包围，所用的精良材料和丰富色彩与屋顶黑色外镀金属包层的砖石台阶形成对比。这两种形式由一个透明的玻璃长廊连在一起，这条长廊顺着房屋北面的界墙把外面院落的轮廓框出。

室内和室外的定义很难辨别出来，透明的双褶门使两者不断发生物质和视觉的联系。建筑的延伸部分同时具有内部和外部的功能及视觉感受。

维达屋的开阔和看似简单的性质后来表现出来复杂和模棱两可性。

建筑延伸部分是从以俄罗斯方块形式出现在场地东侧的部件创造出来的。一块随意放置的俄罗斯方块出人意料地深嵌在原有建筑物的墙内，这块石料给主卧配上一个套房，拾级向下就可以走到建筑延伸部分。

Location
Melbourne, Australia
DESIGNERS
Andrew Maynard Architects
COMPLETION
2008
PHOTOGRAPHERS
Andrew Maynard Architects

项目地点
澳大利亚,墨尔本
设计师
安德鲁·梅纳德建筑师事务所
完工时间
2008
摄影师
安德鲁·梅纳德建筑师事务所

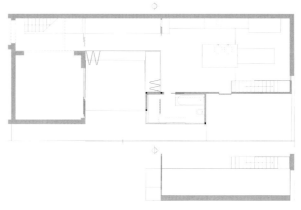

Shuwai

墅外

Located at the 7th floor and the 8th floor of a 20-floor building in Pudong, the apartment features a high-ceiling balcony that connects the two floors. For apartment, it is rare to see the pattern. It is like a mini villa inside an apartment.

At the entrance located on the lower floor, the designers place the staircase at the bottom side as the swivel hinge of the entire spacing. On one hand it is used as a view of the entrance, on the other hand it also performs the function as a segmentation to divide and define the entire spacing into left and right, up and down. The staircase is utilised as the horizontal and vertical space occlusion interface.

On the bottom side of the balcony places a two-floor-high mirror to magnify the entire space. The right hand side of the staircase is the dining-room. It connects to the western kitchen and the Chinese kitchen. The ceiling has an approximately sixty-centimeter hollow to set off partial vision link to the study room. Follow the vertical vector staircase to the private area on the second floor.

本身坐落在浦东一栋20层公寓的7、8层，有一个挑高的露台连接着两个楼层。公寓中这样的格局并不多见。有点像是公寓里的小别墅。

在底层楼层的入口处，我们把楼梯直接放置在底端作为空间整体的旋转枢纽。一方面作为入口端景，另一方面也把通往左右和上下的空间划分界定出来，利用楼梯作为水平和垂直空间的咬合接口。

在阳台底端放置了两层楼高的镜子，放大空间横纵景深。楼梯右边是餐厅，连接了中西厨房。天花板被锯开了一个60厘米左右的开口，用以跟楼上书房做局部的视觉连接。沿着垂直向上的楼梯来到二楼的私密空间。

Location
Shanghai, China
DESIGNERS
Hank M. Chao / MoHen Design Internationa
COMPLETION
2009
PHOTOGRAPHERS
MoHen Design International/Maoder Chou

项目地点
中国.上海
设计师
牧桓设计国际 赵敏
完工时间
2009
摄影师
牧桓设计国际 周宇贤

The World's First Underwater Spa

世界首个水下疗养地

The world's only underwater spa has recently reopened after an extensive redesign by architect Richard Hywel Evans, the man behind the world's first carbon free hotel.

The underwater spa, which forms part of the luxury Huvafen Fushi resort in the Maldives, comprises two double treatment rooms and a separate relaxation area with mind-blowing views under the Indian Ocean. The $180,000 redesign of the interior creates the boldest and most sensory appealing spa in the world.

Guests enter the underwater spa along a passageway lit with colour-change lights in the ceiling to enhance the overall sensory experience. Once inside, reconfigurable sliding walls allow the space to be opened up to make the most of the spectacular views or closed to create a more intimate space for treatments. The transformation also includes lifting the interior wall colour from a light wood to a double curvature organic form, installing a stretch Barisol ceiling and replicating the sea floor by laying blue resin bonded pebble tiles under the foot. The resort itself comprises 43 naturally modern rooms located both on the beach and on stilts over the Indian Ocean.

世界上唯一一个水下疗养地最近重新开放。疗养地的扩建由理查德·海威·埃文斯设计，他也是世界上首家无碳酒店的设计师。

这个水下疗养地也是马尔代夫芙花芬豪华度假胜地的一部分。它由两个双层表皮的房间和一个休闲区组成，在这儿人们可以看到印度洋水下令人兴奋的景色。耗资18万美元的室内设计创造了世界上最大胆的、最吸引人的疗养地。

客人通过通道进入水下疗养地。从入口到内部的通道顶棚是色彩不断变换的彩灯，给人以全新的感官体验。走入里面，可重构的滑动墙可以让空间展开，让游客看到最壮观的景色，也可分隔出更私密的空间。室内墙的颜色还可以从轻质木材转换成双层、弯曲的有机外形。顶棚由Barisol的轻柔的发光材料制成。为了重述海底世界，地面铺了蓝色的仿鹅卵石树脂瓦。度假胜地自己还有43间天然、现代特色的客房，分布在海滩上或高挑于印度洋的上面。

Location
Maldives
DESIGNERS
Richard Hywel Evans Architecture & Design Ltd.
COMPLETION
2009
PHOTOGRAPHERS
Richard Hywel Evans Architecture & Design Ltd.

项目地点
马尔代夫
设计师
里查德·海威·埃文斯建筑设计公司
完工时间
2009
摄影师
里查德·海威·埃文斯建筑设计公司

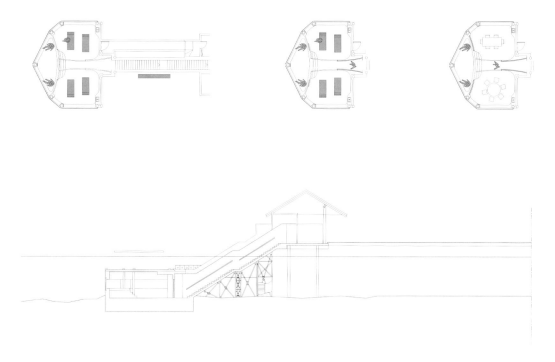

Cidade Jardim Mall

西大德·加帝购物商场

São Paulo is a car driving city; pedestrians' hangout is not privileged. So, this mall project is an attempt to rescue that public street atmosphere. However, with all necessary security and services; it can be seen at the open spaces, in sidewalk floors, in the design of furniture and all the green.

Cidade Jardim Mall is an outdoor mall, such as Ball harbour shops in Miami, but in addition to first-line international brands the new shopping Centre goes further when the subject is the design sophistication, taking care of the project, choosing the material... The main inspiration came from the streets of an idealised city which unfortunately isn't São Paul, well paved streets and well lit, with an ideal thermal comfort and with a lot of Green spaces.

Mainly for being opened and not using very expensive materials that compete with the stores' projects, a low profile-chic Fougeres floor, brown aluminum handrail, ceiling fans and furniture were pecially designed to be in an elegant and safe street atmosphere, exactly the difference just said above from this mall and the other malls in the City.

圣保罗是一个以汽车为主要交通工具的城市。步行街不是很多。因此，该购物商场的设计是想让人们有一个公共空间。并且，这里具备所有必要的安全和服务设施。在这开放的空间里、在每层的通道处，从家具的设计、绿色植物的布置上，四处可见公共街道的氛围。

西大德·加帝购物商场是一个户外商场，如同迈阿密的保港口商店。除了展示一些国际品牌，对设计的思考还进一步深入到设计的各个方面，比如选材。设计师的灵感来源于理想的城市街道设计，但不是圣保罗。这里街道很平坦、很亮堂，温度适宜，有很多的绿色空间。

商场主要是开敞式的设计。和一些商店项目相比，并没有使用很贵重的材料。低调、雅致的富热尔地面铺面、棕色的铝制栏杆、顶棚风扇、家具都特别为这里设计，以创造一个典雅、安全的街道氛围，以区别于城市中其他的商店和购物商场。

Location
São Paulo, Brazil
DESIGNERS
Studio Arthur Casas
COMPLETION
2008
PHOTOGRAPHERS
Studio Arthur Casas

项目地点
巴西,圣保罗
设计师
亚瑟加萨斯设计工作室
完工时间
2008
摄影师
亚瑟加萨斯设计工作室

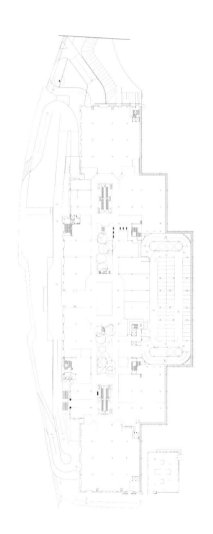

Imall Coloure

怡玛伊考拉沙龙

This beauty parlor is located in the basement of the newly completed "Omote-Sando Hills" complex in posh Harajuku district of Tokyo. "Imai Coloure" is a "total beauty salon" boasting 424 square meters of floor space and is one of the largest shops in the new commercial complex.

The overall style of the design shows the beauty of fresh and clean. The green plants can be seen everywhere. Wooden ceiling, soft lighting, earth-coloured floor tiles, green wall and the pools, all of these create a scene of a summer evening. People like to rest in the shade garden in general. The original Sofa couch gives the comfortable feeling.

"Relaxation" was the driving design concept. To achieve this objective a gently falling waterfall 3 meters tall and 20 meters long with a pool of reflective water was planned as a highlighting feature. The water "wall" consists of corrugated glass backlit to provide a warm, indirect lighting effect. Mirrors of polished stainless steel were erected as free-standing plates that almost disappear in the man-made landscape. Lighting was deliberately kept minimal only to shine on the customers' faces/heads as well as to highlight certain attractions such as the water and the green planted in and out of the pool.

东京繁华的河拉祖库区有一幢新落成的"欧莫特–萨安多山"大厦，怡玛伊考拉美容院就位于这里的地下室。该美容院面积为424平方米，是这个商业大厦中最大的店面之一。

设计的整体风格透着清新自然之美，到处可以看到绿色植物。木制的天棚、柔和的灯光、泥土色的地砖、碧绿色的背景墙和水潭，为客人营造出一个夏日傍晚的场景。客人们如同在田园中休息乘凉一般。造型别致的沙发躺椅更给人以轻松舒适之感。

"娱乐消遣"是该店的设计理念。为达到这个目的，设计了一条3米长的瀑布和一个20米长的水潭。由明亮的纹理玻璃构成的水墙打造温暖气氛的同时，还为空间提供了间接照明。店内竖立的抛光不锈钢镜为独立薄片，在这样的人造风景中几乎看不到了。灯光被刻意调暗，只能照到顾客的面部和头部。在灯光的映衬下，瀑布和绿色植物显得格外美丽。

Location
Tokyo, Japan
DESIGNERS
Edward Suzuki Associates
COMPLETION
2007
PHOTOGRAPHERS
Edward Suzuki Associates

项目地点
日本.东京
设计师
爱德华铃木同仁公司
完工时间
2007
摄影师
爱德华铃木同仁公司

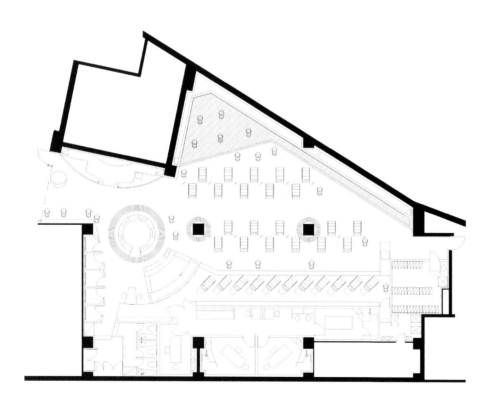

Tribeca

迪比克俱乐部

Tribeca features four distinct areas, ranging from the psychedelic entrance lounge to the ultra chic Manhattan VIP lounge rooms.

The entrance lounge, with plush sofas and psychedelic purple fluorescent lighting leads visitors smoothly to order their drink at the stylish side bar. Designed to encourage mingling, the room features curving couches facing each other, with plenty of room for relaxing while waiting for friends to arrive.

Following the curving walkway, one enters the main bar area. Here orange colours provide a distinct contrast to the entrance's purple lighting, with impressive chandeliers lighting the room. The 360-degree main bar, situated in the centre of the room and adorned with Italian glass mosaic tiles, offers visitors a chance to watch people from any angle.

In Hong Kong, venue's space is always at a premium. Lining the side walls is a unique shelving system that eliminates the need for space taking cocktail tables. Further inside the club, just behind the island bar, is the performance stage. Covered in a similar Italian glass mosaic as the main bar, the stage features a fluorescent lit floor that can change colour to suit the mood. Tribeca also boasts two VIP rooms designed in a Manhattan lounge style.

四大特色区，从颇具迷幻色彩的入口会客厅到高贵典雅的曼哈顿贵宾休息室每一处都独具匠心。

入口会客厅豪华舒适的沙发摆放有致，迷幻的紫色灯光闪烁不停，引领客人走向一侧充满格调的吧台，在这个气派十足的吧台前喝上一杯也别有一番滋味。

蜿蜒的走廊直通迪比克的主吧台。这里枝形吊灯发出的橙色光芒与入口处的紫光色彩对比强烈，给人留下深刻的印象。主吧台建在室内中央，由意大利玻璃马赛克装饰而成，360°可视视角使客人不论坐在哪个位置都可领略它的风采。

香港可谓是寸土如金。沿墙面设计的独特的摆架系统，省去了摆放鸡尾酒桌的空间。沿俱乐部往里走，在岛型吧台后面设有一个舞台。跟主吧台一样，舞台由意大利玻璃马赛克装饰而成。舞台上面的荧光地板别具一格，可随时随景变幻各种颜色。两个仿造曼哈顿休闲风格设计的贵宾休息室是迪比克的又一特色。

Location
Hong Kong, China
DESIGNERS
Mr. Kinney Chan
COMPLETION
2008
PHOTOGRAPHERS
Mr. Kinney Chan

项目地点
中国,香港
设计师
陈德坚
完工时间
2008
摄影师
陈德坚

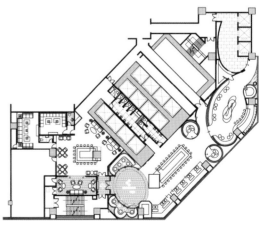

Honda Scooter Store

本田小型摩托车商店

The 100sq.m Chiswick store, selling both bikes and clothing accessories, is housed in a standalone pavilion.

Located in suburban London, the store is full of witty references to life on the street, with a cash-desk clad in green traffic light lenses, a central display feature resembling a traffic island, and perimeter shelving with cut-outs in the shape of road signs.

The central display is illuminated by aluminium lamp stands that echo street lighting, while giant rear-view mirrors mounted on the perimeter walls provide reflected views through the space. A modular window display system features vacuum-formed plastic units which slide along a rail system for flexibility.

Other highlights include ceiling lighting shaped like chevrons, and graphics for windows and walls formed from junction outlines inspired by a popular street atlas. These shapes are combined to create the forms of trees, lamp posts and bollards.

The Chiswick store follows an extensive "blue skies" concept exercise by Jump Studios, which examined how Honda's new "Power of Dreams" communication strategy could influence the design of its scooter retail operations. Jump Studios worked closely with Wieden+Kennedy, the agency behind Honda's UK image overhaul, in order to reflect the brand values of humanity, playfulness, social awareness and optimism.

奇齐克商店面积100平方米，出售自行车和衣着附属品。它是一间独立的屋子，位于伦敦郊区。

商店充满了对街头生活机制进行参照的物品，现金收款台的金属包层是一种绿色信号灯的透镜造型，仿拟交通岛的特点。此外架设栅架的边缘也是按照路标的样子挖空的。

店内中心的展示由铝制灯柱照明，与街灯遥相呼应，安在墙壁边缘的巨大后视镜反光空间。店内标准组件的橱窗展示系统是在轨道上滑动的真空成形的塑料制品，富有弹性。

其他的亮点包括警察V形标志造型的天花板顶灯、窗户的图形和由连结轮廓形成的墙壁，这是受一本流行街头地图册启发得出的灵感。这些结合起来，一起创造出树、灯柱和短柱的形状。

奇齐克商店追求由跳跃工作室实践的"蓝天"的概念，这也检验出本田的新"梦想动力"的表达策略如何影响小型摩托车零售商店的设计。跳跃工作室与威登肯尼迪建筑公司紧密合作，努力反映出本田人性、活泼、社会意识和乐观主义的品牌价值观。

Location
London, Uk
DESIGNERS
Jump Studios
COMPLETION
2007
PHOTOGRAPHERS
Jump Studios

项目地点
英国.伦敦
设计师
跳跃工作室
完工时间
2007
摄影师
跳跃工作室

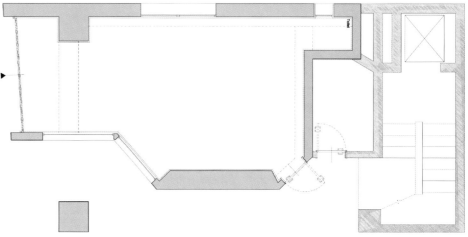

Kelly Hoppen
凯丽·赫本

The personal sense is the roof of inspiation
—— An interview with Kelly Hoppen

以直觉激发灵感
——访凯丽·赫本

Kelly Hoppen is best-known as the interior designer whose calm, elegant aesthetic has permeated our consciousness. Having started her business at the age of only sixteen and half years old, Kelly is able to look back at a vast amount of experience and has established herself as one of the International Interior Designers. Her celebrated interior design studio, which employs 35 members who manage projects around the world, has completed international schemes for houses, apartments, yachts, ski chalets, a sports centre, hotels and numerous corporate spaces, private jets as well as the interiors for British Airway's first-class cabins. Kelly has also successfully implemented her unique approach across a number of business areas, firmly establishing her reputation as designer, retailer, author, educator, and innovator. Her books have been translated into numerous languages and her work has been on the front covers of magazines worldwide.

凯丽·赫本是一位杰出的室内设计师，因其宁静、简洁、优雅并富有创意的设计而闻名。16岁时，凯丽·赫本创建了以自己名字命名的室内设计公司。现在，公司已拥有35名成员，负责国际性项目的运营。其业务范围涵盖住宅、公寓、游艇、度假小屋、体育中心、酒店、写字间、私人飞机等不同的领域，就连大不列颠航空公司飞机的头等舱也是由该公司进行设计的。不仅在室内设计领域表现突出，精力充沛的凯丽·赫本还涉足于多项商业领域。除了设计师，她还是零售商、作家、教师和创新者。她的书已经被译成多种语言，其作品也被登在世界各地杂志的封面上。

HOTEL DESIGNERS SAY: For a Chinese high end design hotel book which is very interested in featuring Murmuri Hotel.

1. As an expert in hotel design, what do you think is the main difference between hotel design and other interior project?

There are many differences between – hotel and residential design for instance. Hotel design takes into consideration the needs of a large number of guests and public spaces, while an individual, couple or family will be of primary importance when considering a residential project.

Hotel design can be themed, take for example the latest trend for Boutique hotels. A wide range of ideas can be fused together to create the look required. You can have a spa hotel catering to guests looking for wonderful treatments, yoga, total relaxation, and you can have a fun family hotel that looks after the entire family with activities for both parents and their children or a city – business hotel catering to professionals who travel on a regular basis but want the comforts of home.

Residential design on the other hand will take into account the needs and wishes of an individual and is completely unique to that client.

2. How can the local features be fused into hotel design?

Local features are very important to hotel design.

Take for example the landscape and weather. If a hotel has been built in a rather warm climate, then light fabrics like linens and cottons are great for soft furnishings.

Likewise a hotel built in the middle of the Arctic will have fantastic furs as throws on couches and beds and luxurious velvets as curtains and upholstery.

If the hotel has been built by the sea, then flooring such as wood with touches of sisal and abaca rugs are best.

A hotel in Morocco for example could look great with mixes of the beautiful colours that are found naturally within the area, as well as traditional sculptures and paintings.

Great hotel design fuses the talent and signature look of the designer with the natural beauty and heritage of the area the hotel has been built in.

3. What usually inspires you most when designing a hotel?

When designing a hotel, I think personal sense to the project is very important. It is necessary for the designers to have a real sense to the whole style of the hotel or the disposal of the details. Therefore, I can say that the personal sense is the root of inspiration.

4. Do you like the fashionable elements and how to employ them in the hotel design?

Fashionable elements in design can be a lot of fun when used in a sophisticated

1. 您作为酒店设计师，请阐述酒店设计同其他类型的室内设计相比，最大的不同是什么？

酒店设计和家居设计就有很多不同之处。

酒店设计要考虑到众多宾客的需求还有公共空间的构造，而家居项目则要将个人、情侣或是家人的需要作为考虑的首要方面。

酒店设计可以有特定主题。例如作为最新流行趋势的精品酒店。大量的奇思妙想融合在一起满足人们的视觉渴望。可以建一座融矿泉疗养一体的酒店，提供舒适的理疗和瑜伽服务，使身体得到充分的放松，以满足向往此类服务的宾客的需要；也可以建一座别有情调的家庭旅馆，在这里父母和孩子一起娱乐，得到放松。当然，还可以建一座商务酒店，经常出差的人在这里则可以享受到家的舒适与温馨。

然而，住宅设计则要考虑到住户的需要和愿望，对于客户而言是独一无二的设计。

2. 如何在设计酒店中融入当地城市风格？

地方特色对于酒店设计来说是非常重要的。

例如，当地的风景和气候会影响到酒店的设计。如果酒店是建在一个非常温暖的地方，室内装饰品选用浅色的亚麻布和棉布这样的织物就再好不过了。

而建在北极中部的酒店则需要选用极好的皮毛做沙发垫和床垫，用奢侈的丝绒作窗帘和装饰材料。

如果酒店是建在海边，铺地板的材料最好选用高光的剑麻和麻蕉材质的地毯。

例如摩洛哥的一家酒店，设计师将当地几种自然的色彩混合搭配，用传统的雕塑和绘画作装饰，使得这家酒店看上去典雅别致。

way. If mixed with classic design it can look wonderful.

From new fashionable colours, fabrics, and shapes to fashionable ideas from the past – such as art nouveau.

Using such elements in a way that appreciates the past and mixes well with the present can create an interior that is simply divine.

5. Can you illustrate some details you are satisfied during the design process?

I love looking at the design before and after of an interior decoration. The transformation from what a room looks like before it has been decorated to the finished look after I have designed it.

I also love the elements of texture that can be achieved by using a wide range of contrasting fabrics and furniture. The attention to detail with lighting and how that can paint a certain mood, and the addition of art / photography to walls.

6. Interior furnishings play more and more important role in hotel design, and what is your perspective?

Absolutely! The right furnishings create the mood and comfort for good hotel design. Without fabulously tactile fabrics and beautifully made furniture, you cannot have a room that is pleasing to look at and comfortable to stay in. You want the guest staying at a hotel to be relaxed and to enjoy their stay. This requires more than great service, a certain ambiance needs to be created, and that is where design plays a big part.

7. In recent years, the green concept has been extensively adopted in interior and what measures have you taken to realise the eco-friendly goal in hotel design?

As an interior designer it is my responsibility to work with suppliers who act responsibly in the manufacture of their products.

Working with suppliers whose manufacturing processes are in direct agreement with ECO concepts, and whose production enhance the world around us, is the responsibility of all of us.

8. What is the tendency of hotel design?

The tendency of hotel design will be diverse. In other words, there will be more elements adding to hotel design. Different design ideas will be mingled

好的酒店设计，不仅融入了设计者的才华，而且把当地的自然风光和传统风格完美地结合在一起。

3. 灵感是创作的源泉，请问您的酒店设计灵感来源何处？

设计酒店的时候，我觉得个人对项目的感觉很重要。不管是酒店整体的风格还是细节的处理，都需要设计师有真实的感觉。因此，我觉得激发我灵感的就是我的直觉。

4. 您在酒店设计中如何结合时尚元素？

时尚元素在设计中运用得当会有意想不到的效果。如果将它与经典设计相结合的话会很妙。

时尚元素包括流行的色彩、织物、形状以及前卫的设计理念，比如新艺术主义。

对过去的事物不是完全丢弃而是抱有欣赏的态度，以这样一种方式运用这些时尚元素，再结合当前的思想，便能够创造出简单、出色的室内设计作品。

5. 请图文并茂地介绍一下您在设计酒店的过程中最满意的一个或几个细节？

我喜欢看设计之前和设计之后的样子。装修前和装修后会是截然不同的。

我也喜欢用大量织物和陈设装饰的反差对比来突出一些有质感的装饰元素。同时还要注意到照明的细节，怎样使其营造出理想的气氛。还有墙上的艺术挂画也很重要。

6. 酒店陈设设计在酒店设计中占有越来越重要的地位，在您的设计中如何突出陈设的作用？

确实是这样的。恰当的陈设为酒店设计营造了舒适温馨的氛围。没有精美的装饰物和漂亮的家具，房间不会那么吸引人，人们待在里面也不会感到舒适和愉悦。要想客人在酒

with each other, which will create a prosperous design field to hotel design.

9. Behind each hotel project, there is always a cultural spirit, how can you interpret it through your design?

By taking into account the history of the area in which the hotel has been built. This will most probably have been stated in the brief. If it is an area of rich traditions in local sculpture or paintings, using these throughout the interior layout could enhance a rich cultural spirit.

10. Nowadays, plenty of fashionable hotels invite architects, brand designer or fashion designer to design for them. What do you think of this phenomenon? Can you talk something about how it will inspire the prospective hotel design?

We live in an age now where designers of any kind can create brands that translate into both interiors and fashion. A famous fashion designer can be asked to design a hotel that reflects the mood he or she is conveying with their fashion.

As long as the ideas are relevant and work, there is no reason this should not continue to work. The future of hotel design will continue to surprise us all.

店得到放松和享受，除了细致体贴的服务外还需要做到很多，比如说需要营造一种氛围，这时设计将起到关键性的作用。

7. 近年来，人们越来越关注绿色环保理念在室内设计中的体现，请问您在酒店设计中采用了哪些绿色环保措施？

作为一名室内设计师，与那些负责的供应商合作是我的责任。

生产过程要符合环保理念，产品本身能够改善我们生活，这是我们大家的责任。

8. 请您谈谈今后酒店设计的趋势将会更集中地朝哪个方向发展？

酒店的设计趋势会更加多元化，不同的设计理念会交织在一起，使得酒店设计成为一个非常繁荣的领域。

9. 酒店设计师如何通过积累多元文化，表现酒店文化的灵魂？

要考虑到酒店所在地的历史。这可能在前面已经简单的谈到过。如果该地在雕刻艺术方面有着丰富历史的话，在室内设计中用到这些元素，那将会大大的提高酒店的文化品位。

10. 现在很多时尚酒店，或者邀请著名建筑师，或者邀请时尚品牌设计师，甚至时装设计师来做设计，您是怎样看待这种现象？对于酒店设计的未来有哪些启示？

我们生活在这样一个时代，任何类型的设计师都可以创建一个品牌，进军室内设计和时尚领域。著名时装设计师可以被邀请来设计酒店，在这里人们可以充分放松并去感受时尚的力量。

只要设计理念符合需要且行之有效的话，没有理由不采用它。未来的酒店设计将会给我们带来更多的惊喜。

Focal Point in a Sea of Colours
——Colourway in Hotels

对焦缤纷色彩
——谈酒店色彩运用

文：谢昕宜

Some designs would not strike you with visual impact on the first sight; however, after observing and tasting for a while, you would be shocked again and again.

Hotel Murano belongs to this kind. The first impression it leaves is that the restoration was done regularly — a bright lobby, and warm and homey rooms. Thrown a second look, however, some peculiarity is to be found.

What strike you most would be the harmonious matches of various colours. Public places, even those with some privacy like hotels, are usually brightened by one single colour; Murano makes a difference. Almost in every space, multi-colour could be seen, without adding a sense of disorder to the overall grand sight.

In order to achieve the harmony with so many colour, the designer cutely popped out a visual focus, as he said, "The backdrop is minimal and neutral to allow the art to be the focus". Comparing the photos before and after the restoration, we will find the focus lies in the added colours in the bar of the lobby. Bright blue backdrop, with strip-patterned glass bar counter, visually further enlarging the blue backdrop; such is the bar. Around this visual focus, grey sofas and colourful cushions make the colours continue, transforming them into the whole space.

The visual focus is not isolated; rather, it is further explored by correspondence and contrast. On the other side of the grey long sofa, blue single-seat sofas provide a second climax in the rhythm of colour, while corresponding with the blue backdrop of the bar, bringing a sense of wholeness and balance. And also the droplights, simple-coloured while sophisticatedly-patterned, contribute to the harmony.

In the design of other spaces, multi-colour is always the first choice. On the other side of the lobby, red is adopted to express hospitality, corresponding with the flames in the andiron and the blinking ornament above. Red sofa in a black background creates a warm atmosphere, while cool-toned cushions make a contrast. The decorations on one side of the corridor are of an eye-catching golden colour, and strips on the grey floor break the dark tediousness. In the bathroom of each room, the space is brightened simply with a flower before the mirror.

The matches of other colours are also designed with visual focuses.

Sizes are also considered; when the background colour is not very bright, colour matches with sharp contrast are strengthened. For example, in Murano rooms, considering with large areas of neutral colours and blue, the three colours of black, red and white are matched with sharp contrast to catch the eye. We would easily focus on these strongly contrastive colours, which somehow strike one as a kind of intense delicacy. Red lights also echo with the sofa.

However, various colours do not necessarily mean visual focus. The glass boats in the lobby are also colourful; for this design, the design team looked to the flourishing local art community for influence, thus bringing a local flavor to the hotel. The colourful boats with little

有一种设计，并非在第一眼就可以给人强烈的视觉冲击，但是在你仔细观察和体会之后，它会带给你一次又一次的震撼。

穆拉诺酒店就是这样，他给人的第一印象是这个酒店翻新工程做的中规中矩——明亮开敞的大堂，亲和舒适的客房。但仔细看去，就会发现这个酒店设计的特别之处。

最让人惊叹的就是多种色彩的和谐搭配。我们所习惯的公共场所——哪怕是酒店这种带有私密性质的场所——通常都只会出现一种色彩提亮整个空间，而穆拉诺酒店则不同，几乎在每一个空间里，我们都可以看到更多的色彩，而且尽管有这么多色彩同时出现，整体视觉感受还丝毫不乱，仍是一派大气和谐。

造就这种色彩和谐的原因之一，是设计师成功的设计了视觉的焦点。如同设计师所说："其背景幕很小，并且为中性色彩，急需让艺术成为此处的焦点。"对比设计前后的照片，在酒店大堂内吧台部分增加的颜色无疑是这个空间的焦点所在。大面积亮蓝色的背景，竖向彩色条纹的玻璃吧台，在成为视觉焦点的同时也起到了在视觉感受上扩大背景的效果。在这个焦点周围，灰色的沙发、彩色的靠垫都成了这个亮色焦点的延续以及向背景空间的过渡。

焦点并不是孤立的，在呼应和反衬下焦点才能在这个空间整体中发挥更多的作用。在灰色长沙发的另一面，蓝色的单人沙发则是色彩节奏设计的次高潮，与吧台的蓝色背景相呼应，带来空间色彩的整体感和均衡感。同样的还有吊灯，色彩简单而造型华丽，是另一种形式的呼应和协调。

在其他空间的设计里，多彩的空间也都有成功的焦点选择，大堂的另一面，用红色带来热烈感受，壁炉里的火焰及上方闪光的装饰成为这一空间的焦点，黑色背景，红色沙发的色彩设计则更好的营造了空间的氛围，冷色调的靠垫恰到好处的点缀了空间；走廊一侧的装饰品有华丽的金黄色，夺人眼球，地

colour deviation didn't provide sharp contrast; on a small scale, they are the visual focus; but in the whole lobby, it serves only as a third climax after the bar.

The match of brightness and darkness also counts for much. Bright blue appearing on a large scale is a peculiarity of Murano. Usually such bright colours are only used as decoration to brighten a space, but that is for designs with simple colours. In colourful Murano, such shining blue is just needed for enlightenment among many contrasting colours. Just because of this bright blue, other colours are in a harmonious match, and just because of the harmonious match, the bright bar doesn't appear abruptly protruding. Such bright colours are only used in the bright lobby; when blue is adopted in the rooms, it is changed into a rather soft background.

Hotel Murano, with its various colours clearly matched as a harmonious whole, is thus featured as a peculiar hotel. Red and blue cover nearly every corner of the hotel as theme colours; particularly the tranquil blue often appears prominently to catch the eye. Comparatively, red is rather "introverted", mostly appear in dark backgrounds to display the warm atmosphere of the hotel. These colours, as a huge background, constitute the theme of the hotel design, providing an indeed shocking visual experience.

板的灰色条纹打破了暗色空间的沉闷；客房的盥洗室里，梳妆镜前的一支小花，就可以让整个空间明快起来。

其他色彩的搭配，依然是围绕着寻找视觉焦点展开的。

有疏与密的搭配，即大色块与小色块的搭配。当背景色不那么夺目的时候，色调对比更为密集的地方会更为突出。比如在穆拉诺酒店的客房里，相对于大面积的中性色和蓝色，红黑白三色通过高密度和强色差吸引人的视线，空间色彩有张有弛，人们的目光自然容易落到更为"精致"的地方。红色的灯饰也起到了很好的陪衬作用。

但是并非凡是色彩密集就一定可以成为视觉中心。大堂中的玻璃船也是密集小色块，这些装饰品因为灵感来源于当地艺术团，所以极富地方特色和民族特色。这些色彩丰富密集而色差不是非常强烈的彩船装饰，在小范围内，玻璃彩船无疑是视觉的中心，而在整个大堂中，它则是继中央吧台之后的次高潮。

明与暗的搭配也是极为重要的，鲜亮的蓝色大面积出现，是穆拉诺酒店设计的又一妙笔。通常的习惯上，如此鲜亮的颜色只用来做提亮空间的点缀之用，但也许那是针对色彩较为简单的设计，在穆拉诺酒店五彩斑斓的酒店里，则需要这样一片亮丽的蓝色。这一片蓝色从众多颜色中跳出来，正是由于这片亮色，才让其他色彩变得和谐，也正是由于其他色彩的陪衬，才让这个蓝色的吧台背景不显得突兀。而这样的颜色也只能使用在大堂这样的开敞空间中。当蓝色延伸入客房时，就化为较为柔和的背景了。

穆拉诺酒店的色彩虽然多，却层次分明，极具整体感，比如红色、蓝色就是蔓延到酒店各个角落的色彩主题，其中较为宁静的蓝色常常鲜亮的出现，夺人眼球，而红色则相对"内敛"，大多数时候只在暗处暗示着这个酒店的热情氛围。他们共同构成酒店色彩设计的主题，形成一个大背景，在时起时伏，时断时续的变化之间创造出惊人的视觉体验。

Peculiar Colour Design
——An analysis of Hotel Riva Design

别样的色彩设计
——河岸酒店设计浅析

文：又清

Red, black and white are the classical colours we are quite familiar with, and are the easiest to be matched with other colours. However, the three colours could also match themselves to provide an extremely shocking visual impact.

Hotel Riva, designed by Jestico + Whiles, is a successful case. The adoption of black, white and highly pure white in large scales helps create a passionate Hotel Riva in the Croatian Island of Hvar, a famous tourist site. Seen as a whole, or from details, the hotel is designed to be fervent and free. The bold combination of passionate red, pure white and cool black set the tune of the hotel to be pure and passionate. Besides, the colours are adopted in large scales, with clear geometrical lines, showing a prominent modern sense. Meanwhile, the repetition of colours in large scales, particularly the black ones, adds a sense of solidness just to the point, effectively avoiding an easy exceeding from passion to impetuousness.

The use of colour in Hotel Riva is quite creative. The designer always adopts colours unexpectedly: the most important part of a hotel — rooms, which usually are designed to be white for a sense of comfort, are boldly made black and red in the bedding by Jestico + Whiles; even the lavabo is chosen to be bright red. Such design is sure to bring an unconventional living experience.

Located in a building of the medieval architecture style, freedom and passion in design are not enough. A good designer is to blend his work into the local atmosphere.

Though freedom is expressed in the design of Hotel Riva, nostalgia is not interfered at all, which mostly lies in the huge black and white images of vintage film stars screen printed onto cotton fabric backdrops to the beds. It's quite possible that you enter a room and surprisingly encounter a film star whom you have been infatuated with, and thus be brought back to the old black and white films and the crazy days of the past. Then that would definitely be a memorable fragment in your journey. The walls with film star images are faintly patterned, as if indicating the trace of time. The faint patterns, together with the black and white images, make you emerged in the memory of the past, of the history of the old building, thus adding a sense of solidness to the passionately coloured interior.

The function of the images on the walls is far beyond that. Being black and white, they don't interfere with the harmonious match of red, black and white. A pair of white pillows on the bed also helps ease the oppression easily made by the dark colours. Meanwhile, the artistic images add some sentimental feeling to the whole room, effectively making one feel closer to the hotel.

And there is the particularly designed fresco with nude figures in the corridor. The fresco, painted on white walls, follows the feature of the hotel of being simple while tensional. Breaking the tediousness, the designer brought us another surprise with the artistic decoration.

Lighting is also designed to be different. The colour of facilities in the outside

红色、黑色、白色，都是我们熟知的经典颜色，也是最常见最易搭配的颜色，不过在他们作为经典搭配色的同时，他们自身的搭配，往往也能产生极具震撼力的效果。

如前所介绍的，J+W事务所的河岸酒店，就是这样一个成功的案例，大量使用黑色、白色和纯度极高的红色，在度假胜地赫瓦尔岛设计出了一个极富激情的河岸酒店。从整体到细节，设计的每一笔，都造就了酒店热烈而自由的特色。红色的热情，白色的纯粹，配上炫酷的黑色，单从大胆的配色，就注定了河岸酒店设计的单纯与恣意，而且这些颜色运用到设计中，都是以大色块出现，干脆利落的几何线条，极具现代感。同时，大色块的重复出现，尤其是黑色色块，恰到好处的增加了整体设计的厚重感，不会因热烈而显得过分浮躁。

色彩设计别出机杼，河岸酒店的设计善于将颜色运用在出乎意料的地方：最为抢眼的客房设计，在大多数酒店都在用白色烘托舒适的同时，J+W则在床上用品的选择上大胆的选择了黑色与红色，甚至于洗手盆都选择了鲜亮的大红色。这样的设计一定会带给酒店的顾客别样的居住感受。

而这个位于中世纪风格老建筑中的酒店，单单有自由与热烈是不够的，一个好的设计师，当然会让他的作品更好的融入到他所在环境中去。

河岸酒店的设计虽然自由，也丝毫不减复古气息。客房的大幅黑白明星照，就能带来些怀旧的感觉。也许会有某一个住客在走入客房的时候，意外邂逅曾经迷恋的影星，随即想起那些黑白老电影和曾经张扬的少年时光，那一定是这次旅行中最为难忘的一个片段。明星照所在的墙壁上，有隐隐的花纹，似乎也暗示着时光的刻痕。同黑白照片一起怀念过去，呼应着这幢建筑裸露的历史，让热烈的色彩霎时间又多了一份深沉，不再浮躁。

黑白明星照的绝妙之处当然不止这些。照片

area can't be seen clearly in the twilight, thus flowerpots with colourful lights are adopted as decoration to give a lively atmosphere. The bar is in bright green — again a bold choice.

If you thought boldness is the only key point in the design of Hotel Riva, however, you are wrong. The designer really did some rational thinking in the design. The corridor is a good example. Different from the red and black in the rooms, corridors are destined to be in light colour for its functional configuration; apparently white walls make us feel broader than black ones. In order to achieve colour continuity, red is adopted in the baseboard, and black in the lines of the fresco on the walls.

Besides, the terrace with a view to the coastline is not decorated with too much red and black; rather, dark red floor and cozy chaise longue easily create a tranquil place for a good rest. Interiors could be seen through windows, where similar design style is followed naturally.

A sentimental sense of freedom is thus created, with rational thinking and common matches of colours. A combination of simple colours gives unexpected visual impact. It is more often than not that the simplest things are to make surprises, just like Hotel Riva in a "red and black" world.

全部是黑白的，没有打乱红白黑搭配的整体感，床头一对白色的枕头也同样缓和了大面积深色调的压抑感。同时，照片的艺术感和线条在大色块中寻求突破，添加更多感性元素，拉近了酒店与住客之间的距离。

同样的还有走廊里设计师精心绘制的裸体人物壁画。壁画延续了酒店设计简单而富有张力的特色，绘制在白色的墙壁上，是打破沉闷的手段，更是增添趣味的妙笔。这些不入俗流的艺术装饰是设计师带给我们的又一惊喜。

灯光设计同样别具一格，暮色中的室外活动区，看不清具体设施的颜色，就用发出五彩光芒的花盆做装点，打造一个活泼欢快的氛围，吧台则用了鲜明的绿色，同样是大胆的手笔。不过如果认为河岸酒店的设计只是取奇取巧，那就错了，在大胆夸张的个性设计中，河岸酒店也做了很多理性的思考。

比如走廊，不同于客房大面积的红黑色调，走廊的特殊空间形态决定了只能选择浅色调，因为白色的墙壁比黑色更能让人感到开阔通透，为了保持色彩的连续性，红色被用在了地脚线上，而黑色只保留在壁画的线条里。

在可以看见海岸线的露台上，也没有运用太多的红色、黑色，只有暗红色的地板和舒适的躺椅，营造出宁静的休息场所。透过窗户可以看见紧连着的室内空间，设计风格自然过渡，也没有太过夸张的设计。

在理性的思考之外创造自由感性的空间，用最常见的色彩打造别样视觉感受，最为简单的色彩搭配，造就了最为夺目的效果。很多时候，往往就是用最简单的东西，才能创造出最为出人意料的效果。如同这个"红与黑"世界里的河岸酒店。

Cross the Border of Design
Mario Wang Marketing Director of Rainbow Dekor

跨越设计的边界
瑞宝壁纸 市场总监 王茂乐

In the field of design, the professional undertake professional work. Profession means the accumulation of experience, thanks to which, errors in design could be avoided as far as possible. Nowadays in the context of culture clash, however, perhaps being professional could only meet the basic need. From numerous successful cases we found that crossover would yield brilliant designs.

The book From Graphic to Space: Interior Crossover Design, systematically introduces the application of crossover in interior design field. Due to the diversity of need for the space we live in, all the design methods we are familiar with could be adopted in space design, for the enjoyment of multi-visual effects and multi-functions.

This design concept just coincides with the cooperation between Rui Bao company and Lan Zi, a contemporary paintress, who crossed from painting to the design of wall paper patterns, trying to elaborate Chinese culture in interior design. Such innovative ideas are welcomed. For the increasingly acute eyes of consumers, we must produce products with comprehensive

在设计界，专业的人做专业的事。专业，意味着资历和经验的积累，因为专业，才能尽可能避免失误的设计。然而，在文化思潮互相冲击的今天，仅仅够专业也许只能满足最基本的需求了。我们通过无数次的成功案例发现，跨界设计更能够碰撞出精彩的火花。

《从平面到空间——室内跨界设计》一书系统的介绍了室内设计领域中的跨界设计的应用。正因为人们所处空间的多样化需求，所有人们熟知的设计方式都可以应用于空间设计，以达到人们可以享有多样化的视觉效果和功能需求。

这种设计理念正如瑞宝壁纸与当代女画家兰子的合作，兰子从画家角色成为家居墙饰的图案设计师，在家居的设计领域里演绎中国的文化。我们欢迎这样的思潮产生，为满足具有越来越高的鉴赏水平的购买者，必须有具备综合优势的产品问世，而跨界设计的理念给了我们这样一个机会，并由此引入更为广阔范畴的艺术于自身产品。艺术家们的思想也得到了更广泛的传播。

在我们的壁纸研发过程中，更深的体会到墙面艺术所传达的不仅是一种图案或符号，传神的艺术品所蕴含的还有人们的精神寄托。这些如果限定了艺术家们可以发挥的领域，将是多么大的损失啊！艺术就贵在能够产生情感共鸣，既是共鸣就可以通过各种方式来传达这种情感。从音乐到绘画，从文字到图

advantages, and crossover design just gave us a timely opportunity, breathing arts in a bigger range of fields into our products. Meanwhile, artists' ideas are more widely spread.

In the process of developing our wall papers, we drawn a profound lesson that wall paper not only displays patterns or signs; instead, the artistic features it contains are to be appreciated carefully. To decline artists from stepping into this field would be a heavy loss. The point of art lies in its resonance, which should be achieved in various ways, from music to painting, from words to pictures, from graphic to space. Designers bring all of them to interior design, to give us more exquisite living styles. Our intention was to invite masters in different fields to merge their concepts and beliefs into wall paper design. In this way, we believe the impact of art could be more incisively impelled.

As one of the most important interior decorations, wall paper not only involves design elements such as pattern, colour, and texture, it also integrates regional culture, contemporary signs, personal beliefs, etc. For example, the wall papers Rui Bao developed — the cloud series (with Chinese character), pleasant sheep series (with contemporary character), not only displays graphic designs in three-dimensional space, but also reinforces visual experience of a glowing vision with light and shadow, achieved through the adoption of a special printing ink.

From Graphic to Space: Interior Crossover Design stands on the frontier of the era, enlightening people who are searching for design in a broader sense. With its guidance, we would be more clear-minded. As the saying goes, "all roads lead to Rome"; we would arrive at an ideal land of art through different ways, appreciating different sceneries on the journey.

We hope that different ideas could be integrated in design field, that inspirations could clash in an active atmosphere, and that spirit could be inherited and developed in an expansive realm. A real master must practice and experience from various fields, to ultimately stand in the forefront for guidance.

片，从平面到空间，室内设计作为设计人所活动的空间，种种设计都应用于此，方组成了我们精彩的生活。我们设想能够邀请到各个领域的艺术大师来将他的理念或信仰融入墙面艺术，必定可以达到更辉煌的艺术效果。

壁纸作为室内设计的重要的装饰手段，不仅融合了图案、色彩、材质的设计内容，还融合地域性文化、时代特征符号、个人信仰等，比如瑞宝开发的祥云系列（中国特征）、喜羊羊系列（时代特征），不仅将平面设计立体展示，还可以让光影设计视觉效果更佳的展示，利用特殊油墨制造出的焕彩视觉。

《从平面到空间——室内跨界设计》在时代的前端，给困苦于局限设计的人们点亮一盏明灯，启示良多。似乎我们的思维更顺达了，正所谓"条条大路通罗马"，我们可以通过不同的渠道和方式，欣赏沿路各异的风景，进而达到共同的艺术的理想境地。

我们希望在设计界，大家能够思想融合，让灵感在活跃的氛围下互相碰撞，让精神在广阔的领域里传承和发展。一代设计大师，必定是在多领域的历练中孕育，进而站在整个时代的前端领航。

Capture refined elegance, create cozy space

By Zhao Jihui
Secretary-general of The Art Display & Decoration Committee of China

营造格致高雅，获取惬意空间
中国陈设艺术专业委员会 秘书长 赵寂蕙

Nowadays, people become more and more aware of the essential role that interior furnishing plays in space design, especially in that it can make space more personal as regards of the artistic aspect, as the concept "Accessory adorning is more important than interior decoration" gradually penetrates. Moreover, some experts from Europe has deemed interior furnishing as the symbol of the "post-interior design era" and paid special attention to it.

The furnishings can be classified into a number of different elements in accordance with their various roles. Generally, they include the following categories: furniture, illuminations, plants, accessories and textiles, etc. Most often, people can choose and combine these furnishings according to their own needs to achieve the desired effect. Though the subject of interior display has evolved and diverged into several types in terms of different styles and genres rendering it the eternal theme in interior design, the elements it consists of will always be the same.

Although the professional skill concerning interior display can be easily acquired through study, it is the designer's cultural cultivation and aesthetic ability which, however is more difficult to get, that affect the final result. Additionally, the space owner, especially his lifestyle, also plays a decisive role in interior display. Not involving the two parts (the designer and the space owner), the professional skill will say nothing, in other words, the space finally designed will be far away from what they want.

This book Interior Furnishing features a large number of excellent images and detailed texts towards various elements to present the glamour of the furnishings. It will make people have more affection to the furnishings and employ them in their room and thus make their living space more colourful and cheerful!

随着"轻装修、重装饰"的室内设计观念的不断深入，人们越来越认识到陈设设计在室内设计中的重要作用，尤其是通过陈设设计可以塑造出空间的个性，增强艺术的氛围。一些欧洲的学者将陈设作为"后室内设计时代"的标志，强调家具、物品的陈设。

陈设物品的划分按照在室内扮演的角色不同可以分成若干要素。常见的陈设要素主要有以下几类：家具、灯具、织物、饰品、生活用品和绿色植物等。可以根据个人的喜好和要求，将这些要素进行组合、摆放来达到不同的效果。陈设设计发展到今天，已经形成了众多的风格和流派，使陈设设计成为了室内设计中的永恒话题。然而无论是哪一种风格或是流派的陈设方式，其包含的要素都是一致的。

陈设专业技能可通过学习容易获取，而指导技能表现的将是设计者深厚的人文底蕴和审美能力，这不是一种能轻易获取的能力，这取决于设计者的生活态度和被设计者的生活方式，设计者与被设计者若不从这两方面出发去指导其技能表现，最终他们将发现得出的陈设空间不像他们想要的，这种空间会与他们的心灵距离不近，总透着"作秀"的成份。

《室内陈设》一书，运用大量翔实的照片和深入浅出的语言来阐释室内陈设的魅力，相信广大读者通过本书的阅读将会更加喜爱室内陈设艺术，并将之运用到自己的室内空间中，从而使生活的环境变得更加丰富而又充满文化的气息。